Project-Based Inquiry Science™

LIVING TOGETHER

Janet L. Kolodner
Joesph S. Krajcik
Daniel C. Edelson
Brian J. Reiser

HERFF JONES EDUCATION DIVISION

IT's ABOUT TIME®

HERFF JONES EDUCATION DIVISION

84 Business Park Drive, Armonk, NY 10504
Phone (914) 273-2233 Fax (914) 273-2227
www.its-about-time.com

President
Tom Laster

Director of Product Development
Barbara Zahm, Ph.D

Creative Director John Nordland	**Production/Studio Manager** Robert Schwalb
Project Development Editor Ruta Demery	**Layout and Design** Kadi Sarv
Editorial Coordinator Sarah V. Gruber	**Production** Sean Campbell / Michael Hortens
Associate Editor Lakiska Flippin	**Creative Artwork** Dennis Falcon
Assistant Editor Rhonda Gordon	**Technical Art** Marie Killoran
Content and Safety Reviewer Edward Robeck	**Photo Research** Carlo Cantavero

ISBN-13: 978-1-58591-603-0

1 2 3 4 5 VH 12 11 10 09 08

This project was supported, in part, by the **National Science Foundation**
under grant nos. 0137807, 0527341, 0639978.
Opinions expressed are those of the authors and not necessarily those of the National Science Foundation.

PBIS **Principal Investigators**

Janet L. Kolodner is a Regents' Professor in the School of Interactive Computing in Georgia Institute of Technology's College of Computing. Since 1978, her research has focused on learning from experience, both in computers and in people. She pioneered the Artificial Intelligence method called *case-based reasoning*, providing a way for computers to solve new problems based on their past experiences. Her book, *Case-Based Reasoning*, synthesizes work across the case-based reasoning research community from its inception to 1993. Since 1994, Dr. Kolodner has focused on the applications and implications of case-based reasoning for education. In her approach to science education, called Learning by Design™ (LBD), students learn science while pursuing design challenges. Dr. Kolodner has investigated how to create a culture of collaboration and rigorous science talk in classrooms, how to use a project challenge to promote focus on science content, and how students learn and develop when classrooms function as learning communities. Currently, Dr. Kolodner is investigating how to help young people come to think of themselves as scientific reasoners. Dr. Kolodner's research results have been widely published, including in *Cognitive Science, Design Studies,* and the *Journal of the Learning Sciences.*

Dr. Kolodner was founding Director of Georgia Techs' EduTech Institute, served as coordinator of Georgia Techs' Cognitive Science program for many years, and is founding Editor in Chief of the *Journal of the Learning Sciences.* She is a founder of the International Society for the Learning Sciences (ISLS), and she served as its first Executive Officer. She is a fellow of the American Association of Artificial Intelligence (AAAI).

Joseph S. Krajcik is a Professor of Science Education and Associate Dean for Research in the School of Education at the University of Michigan. He works with teachers in science classrooms to bring about sustained change by creating classroom environments in which students find solutions to important intellectual questions that subsume essential curriculum standards and use learning technologies as productivity tools. He seeks to discover what students learn in such environments, as well as to explore and find solutions to challenges that teachers face in enacting such complex instruction. Professor Krajcik has authored and co-authored over 100 manuscripts and makes frequent presentations at international, national and regional conferences that focus on his research, as well as presentations that translate research findings into classroom practice. He is a fellow of the American Association for the Advancement of Science and served as president of the National Association for Research in Science Teaching. Dr. Krajcik co-directs the Center for Highly Interactive Classrooms, Curriculum and Computing in Education at the University of Michigan and is a co-principal investigator in the Center for Curriculum Materials in Science and The National Center for Learning and Teaching Nanoscale Science and Engineering. In 2002, Professor Krajcik was honored to receive a Guest Professorship from Beijing Normal University in Beijing, China. In winter 2005, he was the Weston Visiting Professor of Science Education at the Weizmann Institute of Science in Rehovot, Israel.

Daniel C. Edelson is director of the Geographic Data in Education (GEODE) Initiative at Northwestern University where he is an Associate Professor of the Learning Sciences and Computer Science. Trained as a computer and cognitive scientist, Dr. Edelson develops and studies software and curricula that are informed by contemporary research on learning and motivation. Since 1992, Dr. Edelson has directed a series of projects exploring the use of technology as a catalyst for reform in science education and has led the development of a number of software environments for education. These include My World GIS, a geographic information system for inquiry-based learning, and WorldWatcher, a data visualization and analysis system for gridded geographic data, both of which have been recognized by educators for their contributions to Earth science education. Dr. Edelson is the author of the high school environmental science text, *Investigations in Environmental Science: A Case-Based Approach to the Study of Environmental Systems*. Dr. Edelson is currently engaged in research on professional development and implementation support for schools that have adopted *Investigations in Environmental Science*.

Since 1995, he has been the principal investigator on more than a dozen NSF-funded educational research and development grants, and he is a member of the leadership team of the NSF-funded Center for Curriculum Materials in Science. His research has been widely published, including in the *Journal of the Learning Sciences*, the *Journal of Research on Science Teaching*, the *Journal of Geoscience Education*, and *Science Teacher*.

Brian J. Reiser is a Professor of Learning Sciences in the School of Education and Social Policy at Northwestern University. Professor Reiser served as chair of Northwestern's Learning Sciences Ph.D. program from 1993, shortly after its inception, until 2001. His research focuses on the design and enactment of learning environments that support students' inquiry in science, including both science curriculum materials and scaffolded software tools. His research investigates the design of learning environments that scaffold scientific practices, including investigation, argumentation, and explanation; design principles for technology-infused curricula that engage students in inquiry projects; and the teaching practices that support student inquiry.

Professor Reiser also directed BGuILE (Biology Guided Inquiry Learning Environments) to develop software tools for supporting middle school and high school students in analyzing data and constructing explanations with biological data. Reiser is a co-principal investigator in the NSF Center for Curriculum Materials in Science. He recently served as a member of the NRC panel authoring the report *Taking Science to School*. Professor Reiser received his Ph.D. in Cognitive Science from Yale University in 1983.

Acknowledgements

Three research teams contributed to the development of Project-Based Inquiry Science (PBIS): a team at Georgia Institute of Technology headed by Janet L. Kolodner, a team at Northwestern University headed by Daniel Edelson and Brian Reiser, and a team at University of Michigan headed by Joseph Krajcik and Ron Marx. Each of the PBIS units was originally developed by one of these teams and then later revised and edited to be a part of the full three-year middle-school curriculum that became PBIS.

PBIS has its roots in two educational approaches, Project-Based Science and Learning by Design™. Project-Based Science suggests that students should learn science through engaging in the same kinds of inquiry practices scientists use, in the context of scientific problems relevant to their lives and using tools authentic to science. Project-Based Science was originally conceived in the hi-ce Center at University of Michigan, with funding from the National Science Foundation. Learning by Design™ derives from Problem-Based Learning and suggests sequencing, social practices, and reflective activities for promoting learning. It engages students in design practices, including the use of iteration and deliberate reflection. LBD was conceived at Georgia Institute of Technology, with funding from the National Science Foundation, DARPA, and the McDonnell Foundation.

The development of the integrated PBIS curriculum was supported by the National Science Foundation under grants nos. 0137807, 0527341, and 0639978. Any opinions, findings and conclusions, or recommendations expressed in this material are those of the authors and do not necessarily reflect the views of the National Science Foundation.

PBIS Team

Principal Investigator
Janet L. Kolodner

Co-Principal Investigators
Daniel C. Edelson
Joseph S. Krajcik
Brian J. Reiser

NSF Program Officer
Gerhard Salinger

Curriculum Developers
Michael T. Ryan
Mary L. Starr

Teacher's Edition Developers
Rebecca M. Schneider
Mary L. Starr

Literacy Specialist
LeeAnn M. Sutherland

NSF Program Reviewer
Arthur Eisenkraft

Project Coordinator
Juliana Lancaster

External Evaluators
The Learning Partnership
Steven M. McGee
Jennifer Witers

The Georgia Institute of Technology Team

Project Director:
Janet L. Kolodner

Development of PBIS units at the Georgia Institute of Technology was conducted in conjunction with the Learning by Design™ Research group (LBD), Janet L. Kolodner, PI.

Lead Developers, Physical Science:
David Crismond
Michael T. Ryan

Lead Developer, Earth Science:
Paul J. Camp

Assessment and Evaluation:
Barbara Fasse
Daniel Hickey
Jackie Gray
Laura Vandewiele
Jennifer Holbrook

Project Pioneers:
JoAnne Collins
David Crismond
Joanna Fox
Alice Gertzman
Mark Guzdial
Cindy Hmelo-Silver
Douglas Holton
Roland Hubscher
N. Hari Narayanan
Wendy Newstetter
Valery Petrushin
Kathy Politis
Sadhana Puntambekar
David Rector
Janice Young

The Northwestern University Team

Project Directors:
Daniel Edelson
Brian Reiser

Lead Developer, Biology:
David Kanter

Lead Developers, Earth Science:
Jennifer Mundt Leimberer
Darlene Slusher

Development of PBIS units at Northwestern was conducted in conjunction with:

The Center for Learning Technologies in Urban Schools (LeTUS) at Northwestern, and the Chicago Public Schools
Louis Gomez, PI;
Clifton Burgess, PI
for Chicago Public Schools.

The BioQ Collaborative
David Kanter, PI.

The Biology Guided Learning Environments (BGuILE) Project
Brian Reiser, PI.

The Geographic Data in Education (GEODE) Initiative
Daniel Edelson, Director

The Center for Curriculum Materials in Science at Northwestern
Brian Reiser,
Daniel Edelson,
Bruce Sherin, PIs.

The University of Michigan Team

Project Directors:
Joseph Krajcik
Ron Marx

Literacy Specialist:
LeeAnn M. Sutherland

Project Coordinator:
Mary L. Starr

Development of PBIS units at University of Michigan was conducted in conjunction with:

The Center for Learning Technologies in Urban Schools (LeTUS)
Ron Marx, Phyllis Blumenfeld,
Barry Fishman,
Joseph Krajcik,
Elliot Soloway, PIs.

The Detroit Public Schools
Juanita Clay-Chambers
Deborah Peek-Brown

The Center for Highly Interactive Computing in Education (hi-ce)
Ron Marx,
Phyllis Blumenfeld,
Barry Fishman,
Joe Krajcik,
Elliot Soloway,
Elizabeth Moje,
LeeAnn Sutherland, PIs.

Field-Test Teachers

National Field Test
Tamica Andrew
Leslie Baker
Jeanne Bayer
Gretchen Bryant
Boris Consuegra
Daun D'Aversa
Candi DiMauro
Kristie L. Divinski
Donna M. Dowd
Jason Fiorito
Lara Fish
Christine Gleason
Christine Hallerman
Terri L. Hart-Parker
Jennifer Hunn
Rhonda K. Hunter
Jessica Jones
Dawn Kuppersmith
Anthony F. Lawrence
Ann Novak
Rise Orsini
Tracy E. Parham
Cheryl Sgro-Ellis
Debra Tenenbaum
Sara B. Topper
Becky Watts
Debra A. Williams
Ingrid M. Woolfolk
Ping-Jade Yang

**New York City
Field Test**
*Several sequences of PBIS
units have been field
tested in New York City
under the leadership of
Whitney Lukens, Staff
Developer for Region 9,
and Greg Borman, Science
Instructional Specialist,
New York City Department
of Education*

6th Grade
Norman Agard
Tazinmudin Ali
Heather Guthartz
Aniba
Asher Arzonane
Asli Aydin

Joshua Blum
Filomena Borrero
Shareese Blakely
John J. Blaylock
Tsedey Bogale
Zachary Brachio
Thelma Brown
Alicia Browne-Jones
Scott Bullis
Maximo Cabral
Lionel Callender
Matthew Carpenter
Ana Maria Castro
Diane Castro
Anne Chan
Ligia Chiorean
Boris Consuegra
Careen Halton Cooper
Cinnamon Czarnecki
Kristin Decker
Nancy Dejean
Gina DiCicco
Donna Dowd
Lizanne Espina
Joan Ferrato
Matt Finnerty
Jacqueline Flicker
Helen Fludd
Leigh Summers Frey
Helene Friedman-Hager
Diana Gering
Matthew Giles
Lucy Gill
Steven Gladden
Greg Grambo
Carrie Grodin-Vehling
Stephan Joanides
Kathryn Kadei
Paraskevi Karangunis
Cynthia Kerns
Martine Lalanne
Erin Lalor
Jennifer Lerman
Sara Lugert
Whitney Lukens
Dana Martorella
Christine Mazurek
Janine McGeown
Chevelle McKeever
Kevin Meyer
Jennifer Miller

Nicholas Miller
Diana Neligan
Caitlin Van Ness
Marlyn Orque
Eloisa Gelo Ortiz
Gina Papadopoulos
Tim Perez
Albertha Petrochilos
Christopher Poli
Kristina Rodriguez
Nadiesta Sanchez
Annette Schavez
Hilary Sedgwitch
Elissa Seto
Laura Shectman
Audrey Shmuel
Katherine Silva
Ragini Singhal
C. Nicole Smith
Gitangali Sohit
Justin Stein
Thomas Tapia
Eilish Walsh-Lennon
Lisa Wong
Brian Yanek
Cesar Yarleque
David Zaretsky
Colleen Zarinsky

7th Grade
Mayra Amaro
Emmanuel Anastasiou
Cheryl Barnhill
Bryce Cahn
Ligia Chiorean
Ben Colella
Boris Consuegra
Careen Halton Cooper
Elizabeth Derse
Urmilla Dhanraj
Gina DiCicco
Lydia Doubleday
Lizanne Espina
Matt Finnerty
Steven Gladden
Stephanie Goldberg
Nicholas Graham
Robert Hunter
Charlene Joseph
Ketlynne Joseph
Kimberly Kavazanjian

Christine Kennedy
Bakwah Kotung
Lisa Kraker
Anthony Lett
Herb Lippe
Jennifer Lopez
Jill Mastromarino
Kerry McKie
Christie Morgado
Patrick O'Connor
Agnes Ochiagha
Tim Perez
Nadia Piltser
Chris Poli
Carmelo Ruiz
Kim Sanders
Leslie Schiavone
Ileana Solla
Jacqueline Taylor
Purvi Vora
Ester Wiltz
Carla Yuille
Marcy Sexauer Zacchea
Lidan Zhou

Living Together

Living Together was developed by the PBIS development team based on two previously-developed units: *Water Quality*, developed at University of Michigan, and *Can a Bug Save a Farm?* developed at University of Illinois, Urbana-Champaign. *Water Quality* was developed as part of the work of the Center for Learning Technologies in Urban Schools and as a joint project of University of Michigan's Center for Highly Interactive Computing in Education and the Detroit Public Schools Urban Systemic Initiative. *Can a Bug Save a Farm?* was developed as part of the PBIS project.

Living Together

Lead Developers:
Michael T. Ryan
Mary L. Starr

Other Developers:
Francesca Casella

Contributing field-test teachers
Asher Arzonane
Matthew Carpenter
Anne Chan
Lizanne Espina
Enrique Garcia
Steven Gladden
Dani Horowitz
Stephan Joanides
Sunny Kam
Crystal Marsh
Tim Perez
Christopher Poli
Nadiesta Sanchez
Caitlin Van Ness
Cesar Yarleque
Renee Zalewitz

Water Quality

Project Directors:
Joseph Krajcik
Ron Marx

Lead Developer:
Jonathon Singer
Margaret Roy

Other Developers:
Karen Amati
Steven Best
Elena S. Takaki
Valerie Talsma
Rebecca M. Schneider

Detroit Schools Liaison
Deborah Peek-Brown

Can a Bug Save a Farm?

Project Director:
Brian J. Reiser

Lead Developers:
Barbara Hug
M. Elizabeth Gonzalez

Other Developers:
Issam Abi-El-Mona
Jiehae Lee
Heidi Leuszler
Faith Sharp

Pilot teachers:
Bonnie McArthur
Lara Fish

The development of *Living Together* was supported by the National Science Foundation under grants no. 0137807, 0527341, and 0639978. The development of *Water Quality* was supported the Center for Learning Technology in Urban School funded by the National Science Foundation under grant nos. 0830 310 A605. We are indebted to teachers from the Detroit Public Schools for their feedback on the unit. The development of *Can a Bug Save a Farm?* was supported by the National Science Foundation under grant no. 0137807. We are grateful for the recommendations of Whitney Lukens of the NYC Public Schools during development of this unit. Any opinions, findings, and conclusions or recommendations expressed in this material are those of the authors and do not necessarily reflect the views of the National Science Foundation.

Table of Contents

Introducing PBIS

What Do Scientists Do?

1) Scientists...address big challenges and big questions.

You will find many different kinds of big challenges and questions in PBIS units. Some ask you to think about why something is a certain way. Some ask you to think about what causes something to change. Some challenge you to design a solution to a problem. Most of them are about things that can and do happen in the real world.

Understand the Big Challenge or Question

As you get started with each Unit, you will do activities that help you understand the *Big Question* or *Challenge* for that Unit. You will think about what you already know that might help you, and you will identify some of the new things you will need to learn.

Project Board

The *Project Board* helps you keep track of your learning. For each challenge or question, you will use a *Project Board* to keep track of what you know, what you need to learn, and what you are learning. As you learn and gather evidence, you will record that on the *Project Board*. After you have answered each small question or challenge, you will return to the *Project Board* to record how what you've learned helps you answer the *Big Question* or *Challenge*.

Learning Sets

Each Unit is composed of a group of *Learning Sets*, one for each of the smaller questions that needs to be answered to address the big question or challenge. In each *Learning Set*, you will investigate and read to find answers to the *Learning Set's* question. You will also have a chance to share the results of your investigations with your classmates and work together to make sense of what you are learning. As you come to understand answers to the questions on the *Project Board*, you will record those answers and the evidence you've collected that convinces you of what you've learned. At the end of each *Learning Set*, you will apply what you've learned to the big question or challenge.

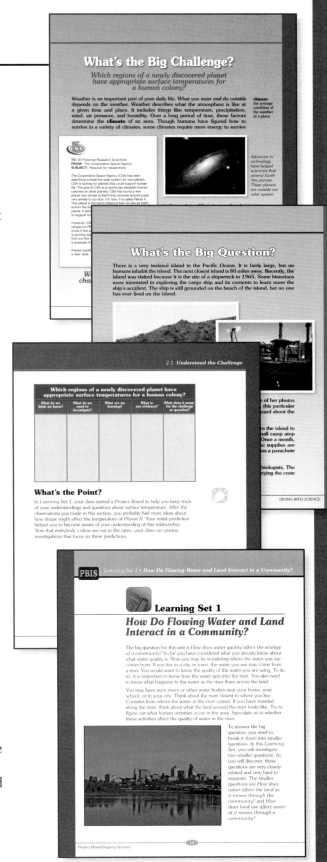

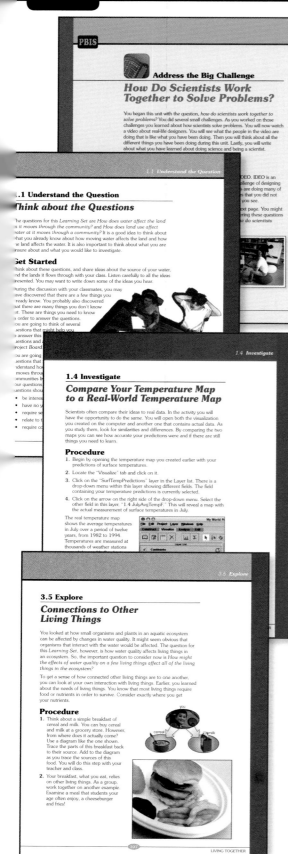

Address the Big Challenge/ Answer the Big Question

At the end of each Unit, you will put everything you have learned together to tackle the *Big Challenge* or *Question*.

2) Scientists...address smaller questions and challenges.

What You Do in a Learning Set

Understanding the Question or Challenge

At the start of each *Learning Set*, you will usually do activities that will help you understand the *Learning Set's* question or challenge and recognize what you already know that can help you answer the question or achieve the challenge. Usually, you will visit the *Project Board* after these activities and record on it the even smaller questions that you need to investigate to answer a *Learning Set's* question.

Investigate/Explore

There are many different kinds of investigations you might do to find answers to questions. In the *Learning Sets* you might

- Design and run experiments
- Design and run simulations
- Design and build models
- Examine large sets of data

Don't worry if you haven't done these things before. The text will provide you with lots of help in designing your investigations and in analyzing your data.

Read

Like scientists, you will also read about the science you are learning. You'll read a little bit before you investigate, but most of the reading you do will be to help you understand what you've experienced or seen in an investigation. Each time you read, the text will include *Stop and Think* questions after the reading. These questions will help you gauge how well you understand what you have read. Usually, the class will discuss the answers to *Stop and Think* questions before going on so that everybody has a chance to make sense of the reading.

Design and Build

When the *Big Challenge* for a Unit asks you to design something, the challenge in a *Learning Set* might also ask you to design something and make it work. Often you will design a part of the thing you will design and build for the big challenge. When a *Learning Set* challenges you to design and build something, you will do several things:

- Identify what questions you need to answer to be successful

- Investigate to find answers to those questions

- Use those answers to plan a good design solution

- Build and test your design

Because designs don't always work the way you want them to, you will usually do a design challenge more than once. Each time through, you will test your design. If your design doesn't work as well as you'd like, you will determine why it is not working and identify other things you need to learn to make it work better. Then you will learn those things and try again.

Explain and Recommend

A big part of what scientists do is explain, or try to make sense of why things happen the way they do. An explanation describes why something is the way it is or behaves the way it does. An explanation is a statement you make built from claims (what you think you know), evidence (from an investigation) that supports the claim, and science knowledge. As they learn, scientists get better at explaining. You'll see that you get better too as you work through the *Learning Sets*.

A recommendation is a special kind of claim—one where you advise somebody about what to do. You will make recommendations and support them with evidence, science knowledge, and explanations.

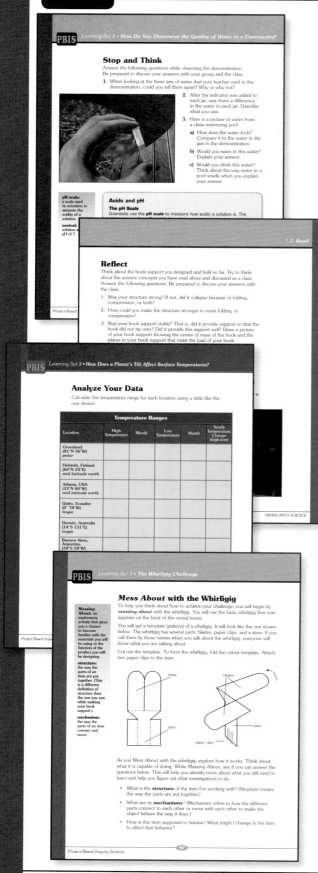

3) Scientists...reflect in many different ways.

PBIS provides guidance to help you think about what you are doing and to recognize what you are learning. Doing this often as you are working will help you be a successful student scientist.

Tools for Making Sense

Stop and Think

Stop and Think sections help you make sense of what you've been doing in the section you are working on. *Stop and Think* sections include a set of questions to help you understand what you've just read or done. Sometimes the questions will remind you of something you need to pay more attention to. Sometimes they will help you connect what you've just read to things you already know. When there is a *Stop and Think* in the text, you will work individually or with a partner to answer the questions, and then the whole class will discuss what you've learned.

Reflect

Reflect sections help you connect what you've just done with other things you've read or done earlier in the Unit (or in another unit). When there is a *Reflect* in the text, you will work individually or with a partner or your small group to answer the questions, and then the whole class will discuss what you've learned. You may be asked to answer *Reflect* questions for homework.

Analyze Your Data

Whenever you have to analyze data, the text will provide hints about how to do that and what to look for.

Mess About

"Messing about" is a term that comes from design. It means exploring the materials you will be using for designing or building something or examining something that works like what you will be designing. Messing about helps you discover new ideas—and it can be a lot of fun. The text will usually give you ideas about things to notice as you are messing about.

What's the Point?

At the end of each *Learning Set*, you will find a summary, called *What's the Point*, of the important things we hope you learned from the *Learning Set*. These summaries can help you remember how what you did and learned is connected to the big challenge or question you are working on.

4) Scientists...collaborate.

Scientists never do all their work alone. They work with other scientists (collaborate) and share their knowledge. PBIS helps you be a student scientist by giving you lots of opportunities for sharing your findings, ideas, and discoveries with others (the way scientists do). You will work together in small groups to investigate, design, explain, and do other things. Sometimes you will work in pairs to figure things out together. You will also have lots of opportunities to share your findings with the rest of your classmates and make sense together of what you are learning.

Investigation Expo

In an *Investigation Expo*, small groups report to the class about an investigation they've done. For each *Investigation Expo*, you will make a poster detailing what you were trying to learn from your investigation, what you did, your data, and your interpretation of your data. The text gives you hints about what to present and what to look for in other groups' presentations. *Investigation Expos* are always followed by discussions about what you've learned and about how to do science well. You may also be asked to write a lab report following an investigation.

Plan Briefing/Solution Briefing/Idea Briefing

Briefings are presentations of work in progress. They give you a chance to get advice from your classmates that can help you move forward. During a *Plan Briefing*, you present your plan to the class. It might be a plan for an experiment or a plan for solving a problem or achieving a challenge. During a *Solution Briefing*, you present your solution in progress and ask the class to help you make your solution better. During an *Idea Briefing*, you present your ideas. You get the best advice from your classmates when you present evidence in support of your plan, solution, or idea. Often, you will prepare a poster to help you make your presentation. Briefings are almost always followed by discussions of what you've learned and how you will move forward.

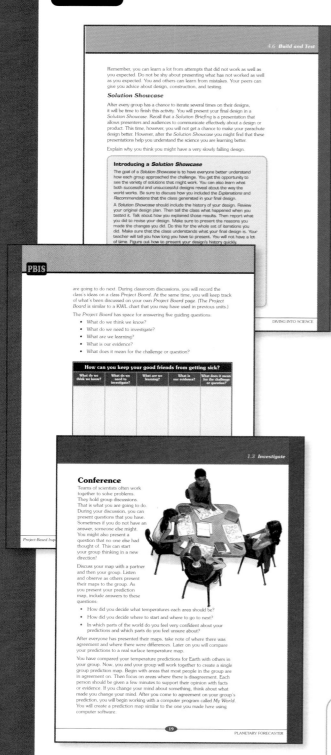

Solution Showcase

Solution Showcases usually appear near the end of a Unit. During a *Solution Showcases*, you show your classmates your finished product—either your answer to a question or your solution to a challenge. You also tell the class why you think it is a good answer or solution, what evidence and science you used to get to your solution, and what you tried along the way before getting to your answer or solution. Sometimes a *Solution Showcases* is followed by a competition. It is almost always followed by a discussion comparing and contrasting the different answers and solutions groups have come up with. You may be asked to write a report or paper following a *Solution Showcases*.

Update the *Project Board*

Remember that the *Project Board* is designed to help the class keep track of what they are learning and their progress towards a Unit's *Big Question* or *Big Challenge*. At the beginning of each Unit, the class creates a *Project Board*, and together you record what you think you know about answering the big question or addressing the *Big Challenge* and what you think you need to investigate further. Near the beginning of each *Learning Set*, the class revisits the *Project Board* and adds new questions and things they think they know to the *Project Board*. At the end of each *Learning Set*, the class again revisits the *Project Board*. This time you record what you have learned, the evidence you've collected, and recommendations you can make about answering the *Big Question* or achieving the *Big Challenge*.

Conference

A *Conference* is a short discussion between a small group of students before a more formal whole-class discussion. Students might discuss predictions and observations, they might try to explain together, they might consult on what they think they know, and so on. Usually a *Conference* is followed by a discussion around the *Project Board*. In these small group discussions, everybody gets a chance to participate.

What's the Point?
Review what you have learned in each *Learning Set*.

Stop and Think
Answer questions that help you understand what you've done in a section.

Communicate
Share your ideas and results with your classmates.

Record
Record your data as you gather it.

Project-Based Inquiry Science

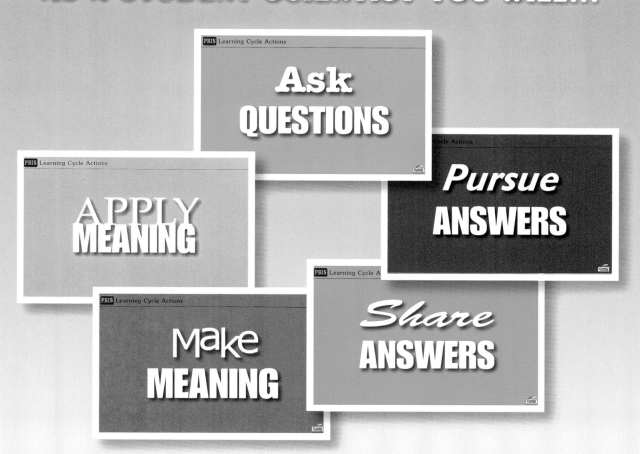

What's the Big Question?

How does water quality affect the ecology of a community?

Water is very important in your life. You drink it, you wash with it, you use it to cook, and you use it for play and exercise. You also know that plants and animals depend on water to stay alive.

If you need water, you can turn on a tap. Towns and cities in the United States have municipal water systems in place. That is where most people get their water. To make sure that the quality of water that you use is good, it is important to know where it comes from.

In this Unit, you are going to investigate how water use affects water quality. You will then look at how water quality affects the plants, animals, and humans in a community. **Ecology** is the study of how plants and animals, including humans, interact with one another and the physical environment.

Look at the *Big Question* for this Unit: *How does water quality affect the ecology of a community?* This is a very big question. To answer the question, you will need to break it down into smaller questions that you can answer. You probably already have some smaller questions that you might want to ask. You will have a chance to ask those questions when you start working on your class *Project Board.*

Welcome to *Living Together.*
This is a great opportunity for you to work as a student scientist.

Think about the Big Question

Before you start, it is a good idea to think about what you might already know about the big question. You will do two activities. They will help you think about how you use water in your daily life. You will also need to think about what is important about the quality of that water.

Get Started: What Is Water Quality?

Your teacher will give your group five jars. Each jar contains a different water sample. With your team, observe the water in the jars. Do not open the jars. Record your observations. Then decide whether or not you would use the water in the jars to fish, swim, boat, or drink. Describe how you arrived at these decisions.

Stop and Think

You have made some decisions about the quality of different water samples.

1. What is meant by quality? What is water quality?

2. How did you determine water quality for the bottles? Was this an adequate method?

3. How else could you measure water quality?

4. If you were walking along a river, lake, or stream, how could you determine the quality of the water?

5. You probably judged the quality of the water from a human's point of view. What if you were a water plant instead of a human? How would you judge the quality of water in each jar?

6. What if you were a fish? Which jar would have good water quality for a fish?

Get Started: What Affects Water Quality?

As you were looking at the samples of water in the jars, you may have wondered where the samples came from and what could make them look so different.

Look at the photos on the next two pages. They are taken in different locations along a river. The river runs through many different types of landscapes and areas. Examine the photos carefully. Think about what the quality of the water in the river might be at each location.

Try to match each water sample in the jar to one of the locations in the photos. Write down which photo you are matching to each jar and why you are making that match.

The river runs through a golf course. Notice the fairways and sand bunkers. In the distance there are several homes.

Here the river runs through a farming community. The rows you see in the dirt were made by the large plows farmers use to plant seed.

Here the river winds through a shipping yard. Notice the docked barges. In the right front there is a plant that produces paint. The paint is then shipped by barge, train, and truck from the plant.

Here the river widens and moves very slowly. In fact, the river enters into a lake at one end, and then it exits the lake through a small stream at the other end. This picture was taken from the yard of a small cottage on the lake. The dock belongs to the cottage owner.

Left: *The river runs past several housing communities and wildlife habitats.*

Below: *At one point, the river is very wide. It is often used for boating. There is even a powerboat race every Fourth of July.*

Above left and right: *This is a small drainage ditch near a highway. This ditch drains into the sewer pipe, and eventually the water flows into a larger part of the river.*

Left: *At the end of the 130-km (80-mile) river, it flows into an even larger river. At this point, there is a large manufacturing plant that makes cars.*

Conference

Share your decisions with the rest of your group. Discuss why you made the matches you did. For some of the jars, you may agree with your group members on the matches you made. For some, you will have disagreements. It is important for each member of your group to discuss why they made their choices. See if you can come to an agreement. Make sure to clearly discuss the reasons for the matches each of you made as you are trying to see if you agree.

This activity may have reminded you of some things that you think you know or don't know about water quality. Jot down notes during the discussion so you will remember what was said when you share again with the class.

Decide as a group what are the most important things to know about water quality and what affects it. What questions are important to investigate to answer the big question?

Your Challenge

Wamego Needs Help!

To answer the big question, you will need to respond to a challenge. A small town requires help in making a decision that will affect its future. Wamego (hwah-MEE-goh) is the small town that needs your advice. It has a population of about 1800. It is located on the banks of the Crystal River. This town has always been a farming community. Most of the farmers grow corn and soybeans. These are the best crops to grow in this area. Nearly 95% of the residents are employed by Wamego's farming businesses. The local economy depends on farming. The other businesses in town all depend on the farmers and their employees (workers). These businesses include a grocery store, gas stations, a movie theater, and several restaurants.

The Crystal River is also important to Wamego. The river is a source of water for the crops. The river is also known as a good trout-fishing river. Trout need very clean, cold water to thrive. Crystal River suits their needs. Every summer Wamego has a Trout Festival. Many people who enjoy fishing travel to the area. The festival celebrates trout fishing and preservation.

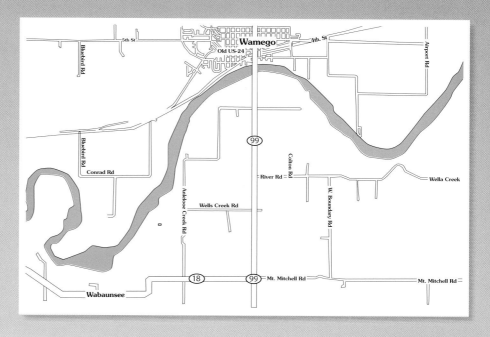

The festival also educates people about what trout need to thrive. The goal of the education effort is to keep the number of trout at a healthy level. In that way, people can enjoy fishing there for many years to come. This festival is fun for many residents and tourists. It is also another income source for the residents of Wamego.

Lately, the farming business has not been good. Crop prices have dropped. The farmers are not making very much money. There is not enough to pay their workers or to support themselves. Some of the farmers have gone bankrupt. As a result, Wamego has lost 15% of its population during the last five years.

The town council is very concerned. They know farming will always be a part of life in Wamego. They do, however, worry about the town losing too many people. They do not want to get so small that there will be very few businesses and residents in Wamego.

FabCo Wants to Move In

A mid-sized manufacturing company called FabCo has contacted the town council. FabCo manufactures cloth. The cloth is sold to companies that make clothes. FabCo is looking for a new location to build its company headquarters and manufacturing plant. FabCo is very interested in relocating to Wamego for several reasons.

- Wamego has a fairly large river and a train line running through the town. This, along with roads, would provide transportation routes for their products.

- The cost of living in the town is low. Their employees would like that.

- The river provides a natural resource (water). Water is important to the production of their cloth.

If FabCo is allowed to move to Wamego, the town could benefit as well. It would mean the following benefits.

- About 15,000 new residents would relocate to Wamego. This would require the building of many new homes, roads, and parks. A new school would need to be built. New businesses offering services to the company and the new residents would be needed. This means more buildings, parking lots, and roads would appear in Wamego.

- FabCo would offer many new jobs to Wamego's residents.

- The town would have money from taxes collected from FabCo and the new residents. This extra money could be used to improve life in Wamego in many ways, including a new hospital.

- The town would not have to depend on farming alone.

Sounds Great! So, What's the Problem?

Many of the residents, including some town council members, are concerned. They worry that FabCo could mean problems for their community. Currently, the land is used for agriculture. If FabCo comes to town, the use of the land will change. The land will be needed for residential, commercial, and industrial purposes. Some people, including the organizers of the Trout Festival, wonder if this will change the river and the wildlife of Wamego.

Wamego residents are not the only ones concerned. Ten miles downstream is the town of St. George. It is also located along the Crystal River.

St. George is an even smaller town than Wamego. It is a resort town. People travel from all over to vacation in St. George. They use the river for recreation. There is fishing, swimming, boating, hiking, and camping in the area. There are several hotels and bed & breakfasts that provide accommodations for tourists. The Crystal River's water quality is very important to St. George's economy and residents. The residents of St. George are worried that the changes in Wamego might affect their lives.

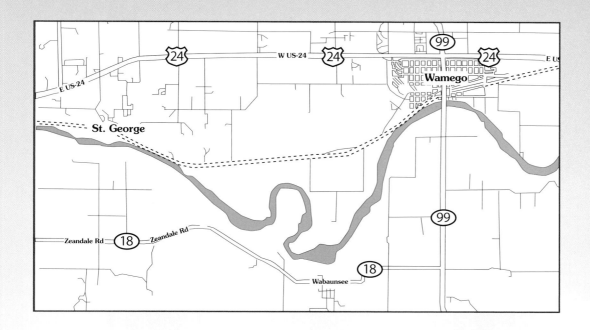

As you answer the big question, you will also take on the challenge of giving advice to the town council of Wamego. What should they take into account in deciding whether or not to let FabCo move in? What will be the ecological advantages of FabCo building its plant in Wamego? What ecological problems might the project cause?

What ecological problems do you think might arise if Fabco moves in? What do you need to learn more about to give the Wamego town council advice? Share your ideas with your group. Discuss the reasons for your ideas. Make lists of what you think might happen and what you think you need to investigate. You will share these with the class when you create a *Project Board*.

Create a *Project Board*

It is useful, when you are working on a design project or trying to answer a hard question or solve a hard problem, to keep track of your progress. You also want to keep track of what you know and what you still need to do. Throughout this Unit, you will be using a *Project Board* to do that. During classroom discussions, your teacher or one of the students will record the class's ideas on a class *Project Board*. At the same time, you will keep track of what has been discussed on your own *Project Board* page.

Recall that a *Project Board* has space for answering five guiding questions:

- What do we think we know?

- What do we need to investigate?

- What are we learning?

- What is our evidence?

- What does it mean for the challenge or question?

To get started on this *Project Board*, you need to identify and record the important science question you need to answer: *How does water quality affect the ecology of a community?* You also need to record your challenge: *What advice should we give Wamego?*

How does water quality affect the ecology of a community? What advice should we give Wamego?				
What do we think we know?	What do we need to investigate?	What are we learning?	What is our evidence?	What does it mean for the challenge or question?

What do we think we know?

In this column of the *Project Board*, you'll record what you think you know about water quality and ecology. Discuss and post the things you think you and your classmates know about water quality and ecology. Have you studied these concepts before? What did you learn then? Even if it is a small fact or idea, talk about it. Discuss any factors that you think might affect water quality, the ecology of a community, and the ecology of Wamego.

What do we need to investigate?

In this column, you will record the things you need to learn to answer the question and address the challenge. During your group conference, you may have found that you and others in your group disagreed about some ideas. You may not know how else to measure water quality. You and your group may not have agreed on where a particular water sample may have been taken. This second column is designed to help you keep track of things that are debatable or unknown, and need to be investigated.

Later in this Unit, you will return to the *Project Board*. For now, work with your classmates and follow your teacher's instructions as you begin filling in its first two columns.

Learning Set 1

How Do Flowing Water and Land Interact in a Community?

The *Big Question* for this Unit is *How does water quality affect the ecology of a community?*. So far you have considered what you already know about what water quality is. Now you may be wondering where the water you use comes from. If you live in a city or town, the water you use may come from a river. You would want to know the quality of the water you are using. To do so, it is important to know how the water gets into the river. You also need to know what happens to the water as the river flows across the land.

You may have seen rivers or other water bodies near your home, your school, or in your city. Think about the river closest to where you live. Consider where the water in the river comes from. If you have traveled along the river, think about what the land around the river looks like. Try to figure out what human activities occur in the area. Speculate as to whether these activities affect the quality of water in the river.

To answer the *Big Question*, you need to break it down into smaller questions. In this *Learning Set*, you will investigate two smaller questions. As you will discover, these questions are very closely related and very hard to separate. The smaller questions are *How does water affect the land as it moves through the community?* and *How does land use affect water at it moves through a community?*.

1.1 Understand the Question

Think about the Questions

The questions for this *Learning Set* are *How does water affect the land as it moves through the community?* and *How does land use affect water at it moves through a community?.* It is a good idea to think about what you already know about how moving water affects the land and how the land affects the water. It is also important to think about what you are unsure about and what you would like to investigate.

Get Started

Think about these questions, and share ideas with your class about the source of your water and the lands it flows through. Listen carefully to all the ideas presented. You may want to write down some of the ideas you hear.

During the discussion with your classmates, you may have discovered that there are a few things you already know. You probably also discovered that there are many things you don't know yet. These are things you need to know to answer the questions. You are going to think of several questions that might help you to answer this *Learning Set*'s questions and add them to the *Project Board*.

With your group, you are going to develop two questions that might help you understand how water changes as it moves through the land where communities live. When you write your questions, keep in mind that your questions should

- be interesting to you,
- require several resources to answer,
- relate to the big question and the river ecology, and
- require collecting and using data.

Also, make sure you avoid yes/no questions and questions with one-sentence answers.

First, develop your own questions. When you have completed your two questions, take the questions back to your small group. Share all the questions with one another. Carefully consider each question and decide if it meets the criteria for a good question. With your group, refine the questions that do not meet the criteria. Choose the two most interesting questions to share now with the class. Give your teacher the rest of the questions so they might be used later.

Update the *Project Board*

How does water quality affect the ecology of the community?				
What do we think we know?	What do we need to investigate?	What are we learning?	What is our evidence?	What does it mean for the challenge or question?

You will now share your group's two questions with your class. Be prepared to support your questions with the criteria you used. Your teacher will help you with the criteria if needed. Then your teacher will add your questions to the *Project Board*. Throughout this *Learning Set*, you will work to answer some of these questions.

Later in this *Learning Set*, you will conduct some investigations and use models to understand how water moves through the land, and how it affects the land it flows through. The investigations will require you to make careful observations and record all your results. The *Project Board* can help you to organize your work as you proceed.

1.2 Investigate

Model How Water Flows In and Around a River

Building **models** is one way that scientists are able to re-create the real world in a lab. Scientists try to represent a phenomenon they are

model: a way of representing something in the world to learn more about.

simulate: to imitate how something happens in the real world by acting it out using a model.

elevation: the height of a geographical location above a reference point, usually sea level.

Scientists use models to simulate processes they can not closely examine in the real world.

investigating. They try to make the model as accurate as possible. In this investigation, you are going to build a model to **simulate** how a river flows. You will use your model to investigate how changes in the areas near the river might alter how the water moves. When you build the model, think about how it is similar or different from a real river.

Design and Build Your Model

Look at the diagram on the next page. Your group will build a model that looks similar to what you see in the diagram. It doesn't have to be exactly what is pictured here. However, you need to have some high spots—areas of high **elevation**. You also need some low spots—areas of low elevation. Discuss with your group where these spots should be.

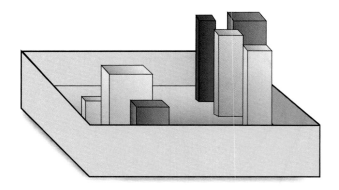

Materials

- **large sheet of butcher paper**
- **building blocks**
- **spray bottle**
- **water**
- **large pan**
- **blue and red permanent markers or crayons**
- **overhead transparency sheet**

Use the building blocks to create areas of higher and lower elevation. Arrange the objects in your tray. Make sure they are spread around and that there is one big object at one end of the tray. Make sure you have a low area running through the middle of your model. This will represent the river.

Crumple a large piece of butcher paper. Be careful not to rip any holes in the paper. Uncrumple the paper.

Carefully cover the objects with the butcher paper by pressing the paper down around them. Use tape to secure the paper to the base of the pan.

Answer the following questions:

1. Which features of a river and the surrounding area do the objects in your model represent?

2. Why did your group decide to arrange the objects the way you did?

Predict

Use a marker. On the butcher paper in your model, write an H on the high parts, and an L on the low parts. Look at your model carefully and observe the way the paper folds and dips around the objects. Think about what would happen if you sprayed water on the paper, as if it were raining. Imagine how the water would move along the paper. Think about where the water would flow. Where would the water pool or puddle? Where would the water drip into the pan?

claim: a statement about what a trend means

On a piece of paper, draw a box to represent the pan used in the model. Write H and L to represent the areas you marked on the butcher paper. Make your sketch as accurate as possible. Use a blue marker or crayon to draw in the paths you think water would take if it rained on your model. Use arrows to show the direction the water will flow.

Draw in the places where the water will pool. Under your picture, write a few words describing what you think will happen when it rains on your model. You can use arrows to connect your words to the different parts of your drawing.

Run Your Model

Now you will simulate how water moves on land when it rains. Use the spray bottle to lightly spray water on your model. The rainfall is supposed to be a light shower. Spray the water carefully so that all parts of your model get wet like they might in a real rainstorm. Everything should get wet equally.

Observe

Observe the way the water flows. When you spray at the top of the highest structure in your model, observe how and where the water flows. Look at where the water stops.

Record your observations as you run your simulation. Use a red marker or crayon. Sketch lines on your drawing of the model representing the actual paths the water took. Mark the places where the water pooled and where the water kept moving. Make your drawings as accurate as possible.

Explain

Consider the blue lines in your drawing that show what you predicted would happen when you added the water to your model. Compare the blue lines to the red lines that you drew to show how the water actually ran. Think about how accurate your predictions were. Did the water flow the way your predicted it would? Describe your results. Use a sentence like this: "I thought the water would flow ….. and when I sprayed water on my model I noticed that the water flowed ….. I think the water flowed the way it did because…" Use a *Create Your Explanation* page. Remember that a good explanation connects your **claim** to your evidence and science knowledge in a logical way.

Make sure that when you spray water on your model, the water goes into the pan, *not* on the floor. Wipe up any spills immediately.

Communicate Your Results

Investigation Expo

Scientists always share their understandings with one another. Presenting their results to others is one of the most important things that scientists do. You will share what you have found in an *Investigation Expo*. To prepare for this, you will use an overhead transparency.

LIVING TOGETHER

Trace the diagram that you drew of your model onto an overhead transparency. Be ready to describe your investigation and clearly detail all your results. The answers to the following questions will be very helpful in preparing your presentation.

- Describe your predictions. (What patterns did you think you would see?)

- Describe why you think your prediction was accurate or inaccurate.

- Describe how the water moved in the model. (What patterns did you see?)

- How did the outcome compare to your prediction? Where did the water flow more quickly? How was the flow different from what you predicted? Did the water pool where you thought it would? How was it different?

- Explain why the water flowed and pooled where it did.

- Describe what you learned about water flow.

As you look at the overheads presented by other students, make sure you can answer these questions. Ask questions that you need answered to understand the results and the explanations others have made.

What's the Point?

In this section, you built a model to simulate how water flows across a landscape when it rains. Placing different-sized objects under the paper created the higher and lower elevations. You also drew a sketch of the model, and you predicted how water would run over the paper. When you ran the simulation you probably noticed that the water always moved from areas of high elevation to areas of lower elevation. Water cannot move uphill. You also noticed that water flowed and created puddles in several places as it flowed. These puddles represent lakes or ponds in the real world.

Water on land works that way too. If you watch where rain falls in one rainstorm, you will be able to predict the path water will take in the next rainstorm. This will be the case as long as the land stays the same from the first rainstorm to the next. But if the land changes, the water will flow and pool differently. You will need to consider how new construction in Wamego might change the land and affect how water flows.

1.3 Read

What is a Watershed?

You just created a model of a river system. You included the river and all the land that drained rainwater into the river. All the water in your model came from rain on the land around the river. The water flowed into the river system. The river and all the land that drains water into the river is called a **watershed**. Everything that sits on the land that drains into a river is *in* the watershed.

For example, imagine a house sitting on top of a hill in the area around a river. That house would be in the watershed. The rain that falls around and on the house, including the driveway and garden, would drain into the river. Water might have to travel very far, but eventually all water ends up in a river or lake.

The diagram shows a watershed. Watersheds come in all shapes and sizes. They cross county, state, and national boundaries. No matter where you are, you are in a watershed!

In a watershed, water flows from higher to lower elevations. The dotted lines in the illustration mark the boundaries of the watershed. The boundaries are in areas of higher elevation. Land on one side of the boundary is in one watershed. Land on the other side of the boundary is in another watershed.

How Does Water Move in a Watershed?

How the water moves in a watershed depends on the shape of the ground. If there are dips in the ground, then water might pool there. When the elevation is high (like the parts marked H on your model), water will run off and move toward areas of the ground that are lower in elevation. The shape of the land determines how fast the water flows. When the **slope** of the land is steep, as shown in the picture at the top of the next page, water runs off very quickly. If the slope is gentler, the water will run off less quickly.

When it rains, water lands on the ground. Then it moves across the ground. The water will continue to run off the ground to collect in lower areas.

watershed: the land area from which water drains into a particular stream, river, or lake.

slope: a measure of steepness. It is the ratio of the change in elevation to the change in horizontal distance.

Scientists call the rainwater that hits the ground and that moves on the ground **runoff**. In your model, you saw a lot of runoff because your paper did not let the water go through.

But not all the water that falls on the ground will run off. Some will run into the ground. The soil can absorb (soak in) some water. This is one place that your model was not a very good model of the river area.

runoff: water from rain or melted snow that moves over the surface of the land.

groundwater: water that is located below the surface of the ground.

The process of absorption is very important. Water that is absorbed into the soil is used by plants and animals that live underground. Once the water is absorbed by the ground, it moves through the ground, always downhill, toward the river. The water that is absorbed by the land and moves under it is called **groundwater**.

Stop and Think

1. In your own words, describe a watershed.

2. In which direction does water flow in a watershed?

3. What is the difference between surface runoff and groundwater?

4. Describe two ways that your model was

   a) the same as a real watershed.

   b) different from a real watershed.

What's the Point?

Watersheds include all the land that surrounds a river. The water in a watershed falls in the watershed and flows to the river. The water can run off the land and flow into the river. This water is called runoff. Sometimes the water is absorbed by the land and flows under and through the soil. This water is called groundwater. Although you cannot see the groundwater, it moves to the river just like the runoff. Plants use the water that is absorbed by the soil.

1.4 Explore

A Case Study: Watersheds in Michigan

You have been looking at how water flows in a model of a watershed. You discovered that water flows from higher to lower elevations.

In this section, you will begin to study a set of real watersheds. The watersheds you will study are in Michigan. By exploring these watersheds, you will see how watersheds connect with and interact with each other. Understanding connections between watersheds will help you give good advice to the Wamego town council. Your teacher is going to show you a type of map of the state of Michigan that shows elevation. It is called a **raised relief map**. It will show you areas of Michigan that are higher and lower in elevation. You will be able to touch the map to feel the different heights. You might think the land in Michigan is flat, but it actually has a variety of elevations. There aren't any large mountains, but there are plenty of large hills. These areas are at a higher elevation than the areas around them.

raised relief map: a three-dimensional map that shows elevations.

Materials

• relief map of Michigan

• topographic map of Michigan

• map of USA

• washable transparency markers

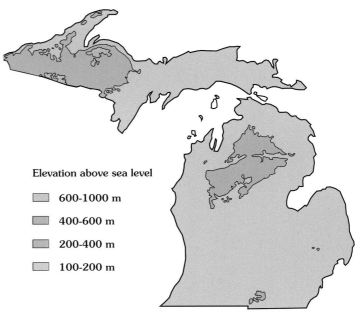

Elevation above sea level

600-1000 m

400-600 m

200-400 m

100-200 m

LIVING TOGETHER

Procedure

1. Compare the raised relief map with a paper map of the same area. The relief map is useful because you can touch it and feel the high and low spots on it. The relief map is a three-dimensional picture of the state of Michigan. The paper map represents the same area shown in the raised relief map. However, the paper map has only two dimensions.

2. Choose one point on your paper map. Compare it to the same spot on the raised relief map. Now look at a high elevation point of the plastic relief map and find the same spot on your paper map. How can you tell on the paper map that this point is an area of high elevation?

3. Use the raised relief map and work in small groups to find one area of Michigan that has a large hill. Starting at the top of the large hill you chose, have one member of your group use a transparency marker to draw on the relief map the direction that water will follow as it runs down the hill. Remember what you learned in the watershed model you built earlier. Water moves downhill.

4. Continue to trace the water path you started to the nearest Great Lake and remember that water cannot run uphill. The members of your group can help the recorder identify the path. If the path seems to be going uphill, you need to find a new path. If you follow a path that is incorrect, wipe off the marker and return to the previous segment of the path. Your challenge is to find a path that does not go uphill and ends up in one of the Great Lakes.

   You have just traced one path that water might take. This might be the path of a river. All the land that drains water into this path is called a watershed.

5. Repeat these steps with each member of your group. Choose a different hill as your starting point each time. Each member of your group should have a chance to draw a water path.

Stop and Think

Answer the following questions. Be prepared to discuss your answers with your group and with the class.

1. How difficult was it to trace a path of water that does not go uphill?

2. Look at the lines you drew to mark the path of water from the top of a hill to one of the Great Lakes. What do these lines tell you about how the elevation of the land in Michigan compares with that of the Great Lakes?

Nested Watersheds

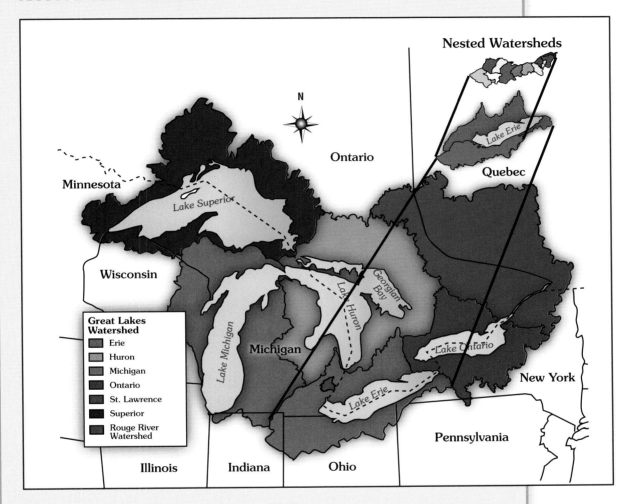

When you traced your paths in the investigation, you might have noticed that there can be some watersheds within other watersheds. Watersheds can be very small or very large. Small watersheds are part of larger watersheds. The creek that might run behind your home perhaps receives only the rainwater that falls in your backyard. That creek has a very small watershed. But it is part of a larger watershed that might drain into a larger creek or a river. This layering of watersheds —small to bigger to biggest—is called nesting. **Nested watersheds** are smaller watersheds that are part of larger watersheds.

nested watershed: one watershed is part of a larger one.

LIVING TOGETHER

The raised relief map of Michigan you used earlier shows you areas of higher and lower elevation. The map also shows you a string of lakes that surround Michigan. This string of lakes touches Michigan on three sides. These lakes are called the Great Lakes. They are the ending point for all the water that runs off the land in Michigan.

All the water on the land in Michigan moves to one of the Great Lakes. The Great Lakes are at a lower elevation than all the land in Michigan. The chart below shows the elevation of each of the Great Lakes. The elevation is measured at the lowest part of the lake. The numbers indicate how many meters above sea level the lakes are.

If you look at the numbers, you will notice that both Lake Michigan and Lake Huron are at a lower elevation than Lake Superior. So, because water runs downhill, water from Lake Superior enters either Lake Michigan or Lake Huron. Water that moves to Lake Huron continues through rivers and canals to the Lakes it is connected to that are lower in elevation—Lake Erie and on to Lake Ontario. The water flows through rivers from Lake Ontario on to the Atlantic Ocean. Water that moves to Lake Michigan flows down the Mississippi and eventually ends up in the Gulf of Mexico.

Elevations of the Great Lakes	
Great Lake	Elevation above sea level
Lake Superior	183 m
Lake Michigan	177 m
Lake Huron	177 m
Lake Erie	174 m
Lake Ontario	75 m

What's the Point?

From your work with your watershed model, you discovered that water always flows from higher to lower elevations. You used this knowledge to trace the path of water from higher elevations to lower ones on a map of a real watershed. You also discovered that there are often many smaller watersheds nested in a larger watershed.

The watershed you explored was in the state of Michigan. You are using a map of Michigan because you are going to study a river in Michigan as a model of how the water quality in a river affects the ecology of the river and its watershed. This will help you understand how the water quality in Crystal River could affect the ecology in its watershed.

Water always flows from higher to lower elevations.

1.5 Read

Introducing the Rouge River Watershed

The Rouge River running through a city.

You used a relief map of the state of Michigan to see how different watersheds nested within one another. You applied what you learned about how water flows from higher to lower elevations to identify the different watersheds in the area. Throughout the rest of this Unit, you will be looking at one specific watershed in Michigan. You will examine the Rouge River watershed. (Rouge is the French word for red.) You can apply what you discover about the Rouge River to investigate other watersheds, including your own.

Where is the Rouge River Watershed Located?

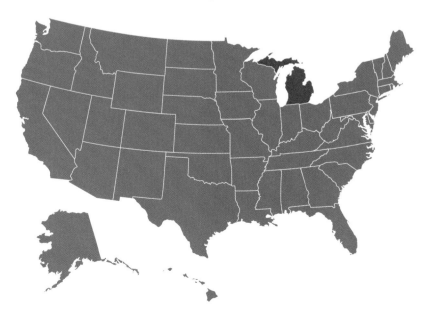

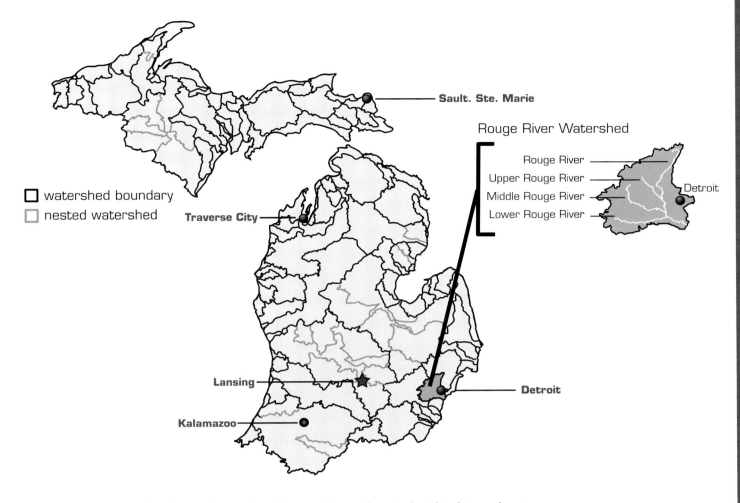

watershed boundary
nested watershed

Rouge River Watershed
Rouge River
Upper Rouge River
Middle Rouge River
Lower Rouge River
Detroit

Sault. Ste. Marie
Traverse City
Lansing
Kalamazoo
Detroit

When you are thinking about the Rouge River, begin by thinking about where it is located in the U.S.A. It is in Michigan.

Next, look at the map of Michigan. The Rouge River watershed is near Detroit. This is an area where many people live.

When you look in the southeast corner of Michigan, you will see that the Rouge River watershed has been highlighted on the map above. The Rouge River watershed is nested in many watersheds in Michigan. It is part of the larger watershed of the Detroit River.

The Detroit River flows into Lake Erie and is part of the Lake Erie watershed. The Lake Erie watershed is a smaller part of the St. Lawrence/ Great Lakes watershed. The St. Lawrence/Great Lakes watershed is part of the huge watershed that covers the entire eastern half of the United States and some of Canada. All these watersheds are nested, one within the other. When water is added to a bigger one, it moves to the smaller ones. Eventually, all the watersheds in Michigan drain into one of the Great Lakes.

How Have People Used the Rouge River?

About 1.3 million people live, work, and play in the Rouge River watershed. Although the river with all its branches is 203 km (126 miles) long, that is still a lot of people. Throughout time, people have used the river for many different purposes. It has supplied drinking water and fish. It has been used for generating electricity. And it has been used as a dumping ground for waste.

Over 150 years ago, there were very few people living in the Rouge River watershed. Most of them lived in Detroit. The Rouge River was clean and had lots of fish in it, even though waste was dumped into it.

Then, about 100 years ago, the population in the watershed started to grow. Many factories for building cars, trucks, and airplanes moved into the area. With the factories came the people who worked in them.

Early industry along the Rouge River in Detroit introduced pollution that severely affected the quality of water.

With the increase in population, the demand to use the river water also increased. More cities and towns began to dump sewage into the river. More river water was needed to generate electricity. Some of the industrial waste was dumped into the river or buried in nearby areas.

Over time, the amount of sewage in the river made recreational activities impossible. Starting around 1950, people began to recognize that the quality of the river was very poor, and something had to be done. In the following years, sewage treatment plants were built. Cities were told they could no longer dump their waste into the river. Areas close to the river were set aside and protected as public parks. Although these measures helped to improve the river environment, illegal dumping of garbage and waste continued. In the 1980s, groups of citizen volunteers decided to remove all the garbage still found along the river banks. These groups have been working on this cleanup ever since. Every year they meet for at least one day to clean up the areas along the river.

Reflect

Look back at the pictures you saw at the beginning of this Unit. These pictures showed many different ways people use a river. All these pictures were taken along the Rouge River, in Michigan. Look at each picture and answer the following questions. Be prepared to share your answers with your group and the class.

1. Does the river flow through a city or town, or through a farm? Can you tell by looking at the pictures?

2. Where do you think the different pictures you saw were taken? Where do you think the industrial parts of the river might be? On a map of the watershed, color that area blue. Where do you think the river is used for recreation? Color that area red.

3. Compare your answers to the answers of other members of your group. Did you all pick the same areas? What else do you need to know to decide if the areas of the map are correctly identified?

4. How do you think the changes in the quality of the river water affect the lives of people living in the watershed?

5. List three examples of human activities that affect the quality of a river. Use what you learned about the Rouge River to explain what might happen to the river, and how the people in the watershed are affected by the changes in water quality.

6. Do you know of groups that are involved in cleaning up the river nearest your home? Do a little research to see if there is a group near you.

Human activities do not always damage the water quality of a river.

Apply

Think about the land around Crystal River. How will it be different with buildings on it? When people build buildings, they cover the land and change the types of land cover that originally existed there. Buildings, houses, and roads all cover the ground. Trees, grass, and soil are moved and may or may not cover the ground the same way they did before. Water can soak into soil, but it flows as runoff when it lands on buildings and roads.

Think back to when you have been outside when it was raining. What happened to the rain when it hit the ground? How did the type of land cover affect what happened to the rain? What kinds of land cover are there in your community?

Think about the watershed you built. You decided that as long as you did not change the structures in the watershed, the water would continue to run off in the same pattern. The direction of the water flow and the places where the water pooled would not change as long as you did not change the objects or the paper. If you had used soil instead of paper, you would have noticed that some water soaked into the ground.

Now you are going to consider the differences among four different land uses. During your class discussion, use a table like the one below to organize what you are learning about each land use.

Residential	Commercial	Industrial	Agricultural

What's the Point?

In the United States, many rivers have been used by people for dumping waste, draining storm water, recreation, and industry. The Rouge River is one example of such a river. There are many other rivers across the United States that have been used in the same way. When people use a natural resource like a river, they modify the environment around it. Human activities in any part of the watershed impact the quality of the river because all the water in the watershed sooner or later ends up in a river. Sometimes water travels many miles, over and under the land, to get to the river. It can carry with it many different materials that can make the river unsafe or unclean.

1.6 Investigate

How Does the River Affect the Land and How Does Land Use Affect the River?

You have been learning about watersheds and how land use can affect the quality of water in a watershed. Now you will look at how water can affect the land that it moves across. Then you will consider the many ways that people use the land in a watershed. You will build your own models of four of these uses: residential, commercial, industrial, and agricultural. Each of these uses changes the land in a different way.

First you will look at a demonstration of a model. Your teacher has built a river model for you to observe. In the model, your teacher is going to use some soil and water to demonstrate how a river might change the land around it. The model will be run several different ways. Each different model will demonstrate a different way for the water to flow down the river.

Demonstration

A stream table can be used to simulate processes occurring in an active river or stream. Since it is not always possible to study these processes in a real

river, scientists use stream tables in the laboratory to do that. They build a model of the river or stream and the land around it. Then they move water through it to simulate water processes. This helps them understand the causes and effects of water processes.

A simple stream table consists of a large pan covered with Earth's material, such as sand, rocks, or grass. A line is drawn in the sand to represent the river. Water drips into the sand from a container at one end. The stream table may be slanted to help the water flow downhill. At the other end of the pan is a drain hole, connected to a bucket, to collect the water.

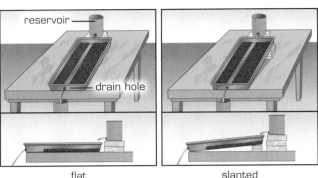

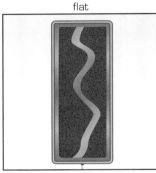 

Natural processes, such as flooding, can change the structure of land and water in a watershed.

Your teacher will set up the stream table in four different ways, as shown in the diagrams.

Sketch the different models. As you watch the water flow through the model, pay very close attention to the way the land on both sides of the river changes. Pay attention to

- how the soil moves,
- where along the bank the soil moves, and
- where the soil ends up.

Make notes about what you observe for each of these situations. You might want to mark your sketches based on what you observed.

Stop and Think

Look at your sketches and the notes you took about the river models you observed. What did you notice about how the soil was moved by the river? Answer these questions. Be prepared to discuss your answers with your group and the class.

1. When the river was straight and the pan was level, how did the soil move along the river?

2. When your teacher made the pan more slanted by lifting the water end of the pan, how did the water move compared to the level pan? How did that change affect the soil that the river moved?

3. Your teacher also made rivers that were more curved. How did that change the way the soil moved along the river?

Erosion and Deposition

Before you start building stream models, you need to know something about how water moves in a watershed and the effect that has on the stream or river. During a heavy rainfall, water in a stream flows fast and will pick up a lot of dirt. As water flows against the bottom and sides of the river channel, it removes more dirt, sand, soil, and debris. Scientists call the "removal of dirt" **erosion**. When water slows down, the collected dirt in the river drops out of the water. Scientists call the debris and soil "dropping out" of water **deposition**.

erosion: a process in which Earth's materials are loosened and removed.

deposition: the setting down of Earth's materials in another area.

land use: how people use Earth's surface.

Design and Build Your Model

When people settle in areas near a river, they change the land around it. Trees, grass, and soil are removed to make room for buildings, houses, and roads. Other areas are used for farming or to grow crops. The way people use Earth's surface is called **land use**. Your class will model the effects of changes in land use on river processes.

Your teacher will assign your group one of the four types of land use to model: residential, commercial, industrial, and agricultural. Read the section describing the land-use model assigned to your group. Use the pictures included in each section to help you imagine the land use that you will model.

You will use a stream table to model effects on the watershed. Begin by building a model as close as possible to one of the models your teacher built. All the groups should begin with a similar model.

After you complete your investigation, you will share the data your group collected with other groups. This way you can compare the effects each land use produces in the river. It is important to be careful when building your model and collecting data so you can compare your results with those of other groups later.

When you build your model in the stream table and watch how water flows through it, pay attention to the areas of erosion and deposition you can see. On a diagram of the stream table, draw and label areas of erosion and deposition. If you are not sure where you are seeing this happen, ask your teacher to help.

Model Residential Use

Suppose a developer decides to build houses next to the river. The land she builds on is covered with grass. The builder builds five houses, equally spaced from each other. Each house includes a two-car garage. Sidewalks connect one house to the other. The builder removes all the soil and grass that was originally on the site. She plans to add a large grass lawn to each house once they are built.

Model Commercial Use

Some land along the river has been designated as a space for building a small mall. The mall will include a movie rental store, a grocery store, a coffee shop, a plant nursery, and a day care. The mall needs 100 parking spaces and will take up the area of about ten houses.

Model Industrial Use

Many industrial factories are built along rivers. The water from the river is often used in the factories for manufacturing and then returned to the river. In your model you are going to build a paper mill. Large trees are brought to your factory to be ground and made into the white paper you write on. The trees are very big. You need to build a road wide enough for trucks to get to the factory in your model.

The factory building will take the area of about ten houses. The pictures shown may help you to imagine how an industrial land use might look.

Model Agricultural Use

You might not be familiar with how land looks when it is used for farming or raising animals. You can use the pictures to help you imagine what this type of land looks like. Build your stream table to look similar to a farm. On your farm, there are a few buildings and a house. The crops have just been planted for the coming year. The model you will build should have a lot of exposed soil. There is a river running through the area. There is also a lake nearby. Families living near the farm use the lake for swimming.

Procedure: Build Your Model

Make a list of the way the land in your model will be used. Be sure to include all the details from the description in your list. Each member of your group should sketch how the land is going to look after all buildings and features of your land use are built. Remember to include all the details from the list you created.

When all the members of your group have completed their drawings, compare them to one another and allow each group member time to describe their plans. Listen carefully for details you may not have included in your drawing.

Now you are ready to build your model. Your teacher will provide you with materials to use. Use your group's drawings to build a model representing the land use you have been assigned.

Predict

Before you try your model, predict how water will flow in your land-use model when you spray water on the different types of land cover. Make a prediction about how water will move on each of the land covers you built in your model. Even if you are not sure, try to guess based on what you already know about that land cover.

As you make your predictions, think about the following questions:

- What will happen to the water that hits the grassy areas? What will happen to the water that hits the sandy areas?

- What makes the water move differently through the different land covers in your model?

- How will the amount of runoff vary among the different land covers?

- How will the time the water takes to get through the different land covers and to the river vary?

- What will the water look like when it enters the river compared to when it fell as rain?

- How clean will the water be that drains through the different land covers?

Procedure: Run Your Model

Spray water on your stream table. Observe where the water flows as it moves through the model.

Recording Your Observations

As you observe how the water flows, record your observations on a data chart. You might also want to record notes on a sketch of your model. Be as specific as possible. Things you might want to check could be: how much of the water you spray ends up in the river; how fast the water flows over the areas where the land cover has changed; does the water form puddles anywhere? Add your own ideas for other interesting patterns you observe.

Analyze Your Results

1. What makes the water move differently through the different land covers in your model?

2. Is the time the water takes to get to the river different in your model compared to the models your teacher demonstrated?

3. How does the amount of runoff vary among the different land covers?

4. How clean is the water that drains through the different land covers?

5. Is the groundwater that enters the model river the same across the length of your model or does it vary?

Communicate Your Results

Investigation Expo

Use the *Analyze Your Results* questions as a way to discuss the results of your investigation in your group.

For the *Investigation Expo*, create a poster with a diagram of your land-use model. Make your diagram as detailed as you possibly can. Include all your land covers as well as your results. Indicate on your diagram places of erosion and deposition, and places where there was a lot of runoff in your model.

During the *Investigation Expo*, you are going to describe to your class how your model worked. You need to include enough details in your presentation so that your classmates will understand how the land cover in your model changed how the water moved. Answer the following questions in your presentation:

- How did the water move in different parts of the stream table?

- How do you think the land cover you modeled might affect how the water is absorbed by the ground compared to vegetation (plant life) or bare soil?

As you listen to the presentations of the other groups, observe how water flows for each land use. Compare the places were erosion and deposition occur in the different models. Compare the amount of runoff produced by different land covers. How do the residential, commercial, industrial, and agricultural land use each affect the amount of runoff produced, compared to that of the bare soil in your teacher's model?

Explain

As the water flowed through the land, you probably noticed that it flowed differently when it was covered by pavement (small laminated pieces), houses, or vegetation. When people have not changed land, water flows in predictable ways. It may flow over the land and run off into a river. If there is more vegetation, the water may be absorbed by the land, be used by the plants, or run through the land as groundwater. No matter how it moves, the water always ends up in the river. When the land use is changed, the patterns of water movement are also changed. More water may run off. The land may absorb less water because it is covered. That could have consequences for plants because they need water at their roots. When water is not absorbed by the ground, it cannot reach the roots of the plants.

1. Think about the stream tables. What is one variable that might affect the amount of deposition in a river? Explain how that variable affects the amount of deposition in the river.

2. What is one variable that might affect the amount of erosion in a river? Explain how that variable affects the amount of erosion in the river.

Explain

Use *Create Your Explanation* pages to help you organize your claims and evidence as you develop your explanations.

Communicate

Share Your Explanation

Your class will meet to discuss each group's explanation. As a class, select or create a set of explanations that explains what affects the deposition in a river and another set that explains what affects the amount of erosion around the river.

What's the Point?

You have built and observed models of a stream with different land uses. As water flowed through your model, you observed how the water changed the shape of the land, and where erosion and deposition were occurring. The movement of soil is critical to how rivers work. The amount of eroded soil carried by the water and the places where it is deposited can greatly affect the balance of a river system.

In addition to changing the pattern of erosion and deposition, different land uses also change water quality. It is important to understand how human land use changes the river and the water so people can look for ways to minimize its impact.

When you changed the land cover in your model, the patterns of erosion and deposition changed. How people use the land around the river can change the water quality.

1.7 Read

More about the Effects of Land Use on a River

Erosion, deposition, and runoff happen naturally. Often, they do not cause problems in rivers. But the changes people make to land affect erosion, deposition, and runoff.

In class, you observed and experimented with a land-use model. The soil represented the land. The stream you made with your finger represented a river. As you sprayed water onto your model, you observed how the water flowed over and through the land. As the water moved, it changed the land.

As you built your model, you changed the way the surface looked. Houses and roads covered what was soil or grass before. You explored the effects of these changes on your model river. You looked at how the land use affected erosion and deposition as well as runoff.

In this section, you will read about ways in which humans have changed the land and examples of how that can affect the nearby rivers. This will help you understand the ways that land use changes the water that ends up in the river. As you read these descriptions, pay particular attention to the land uses you did not model. You will be able to discuss each of the land uses with your class.

Effects of Residential and Commercial Land Use

Both residential and commercial land use affect the river in similar ways. When you built your model for residential or commercial land use, you might have used plastic sheets to represent paved surfaces or roads. You also used plastic blocks to represent buildings. When you ran the model, you probably noticed that water was running off the plastic surfaces. The runoff eventually ended up in the river.

What you saw in your model is very similar to what happens in the real world. Rain runs off the surfaces of buildings, rooftops, and roads. The water falling on paved roads and concrete cannot penetrate the ground. Instead, it flows fast towards the river, increasing erosion. Because the water cannot penetrate into the ground, it is not available as groundwater for plants and trees that need it. Plants and trees may suffer from lack of water. The intense runoff muddies the water in the river because fast-flowing waters can pick up and carry a lot of dirt along the way.

When you created grassy areas in your model, you also noticed that the water did not run off these surfaces. The water was absorbed by the grass and went into the ground. In a real watershed, the vegetation covering the ground acts like a sponge. It absorbs the rain and snow falling to the ground and releases the water over time. This way, vegetation prevents erosion. Because the water it absorbs is released slowly, the dirt in the water is deposited along the way before the water reaches the river. The water entering the river is therefore much cleaner.

Effects of Industrial Land Use

Large industrial factories are often built along rivers. This is because many factories need large quantities of water during **manufacturing** processes. Rivers can provide the water they need. The river can also be used to transport goods and materials in and out of the factory by boat.

manufacturing: the making or producing of anything.

When modeling industrial land use, you used plastic sheets to represent roads and plastic blocks to represent buildings. The effects that these surfaces have on erosion, deposition, and runoff are very similar to what you observed for residential and commercial land uses.

But industrial uses are different in several ways from other land uses. Factories often draw water from the river to run their processes. This water is then returned to the river at the end of the manufacturing cycle. During the process, water can pick up substances or change in other ways. Which changes occur depend on the type of factory and how the water is used. The water exiting a factory can be very different from the water entering one. When this water is discharged back into the river, the changes that occur can be harmful to aquatic organisms or wildlife in the area.

Effects of Agricultural Land Use

Today, many large areas of land are still being cleared to plant fruits and vegetables. As trees and grasses are removed to make room for farms and houses, the soil is exposed. Without the roots of the plants to anchor it, these soils can be eroded away easily. The eroded soil particles carried by the water and discharged into the river affect the water's quality.

The increased erosion of soils used to grow crops does not happen only once, when the land is cleared. When crops are harvested, the soils become exposed again until the next crop grows. More erosion occurs every time the soil becomes exposed. Over time, this increased erosion can change dramatically how a river system works. In some places, the river might widen. In other places, the depositions might change the river's shape or depth.

Soil and dirt from farmlands are not the only things that get washed into the river from agricultural use. Substances added to the soil to help the crops grow and keep insects away are also carried by the water into the river. Some substances are carried in runoff. Others are absorbed into the soil and carried underground in groundwater.

Reflect

Write a description of what happens to the water when it rains in your neighborhood. In your writing you should answer the following questions:

- Where does the rainwater go once it hits the ground?
- Where do you see that the rain is absorbed?
- Where does runoff occur?
- Where do you think most of the water ends up?

What's the Point?

You read about the specific problems different land uses can create. Sometimes the uses create similar problems. The residential and commercial land uses are similar. However, industrial uses are very different from residential uses. When considering the differences between land uses, it is important to pay close attention to how the water is being used and how the land is being changed.

Runoff can carry different substances into water. Parking lots and streets contribute oils, grease, and dirt to runoff. These substances eventually make it into the river.

Erosion and deposition are always issues in land use. The movement of soil is critical to how rivers work. Water can carry soil and deposit it in other places. When soil enters the river, it can affect the quality of the water.

Groundwater is also critical because the water can carry different types of materials through the ground and into the river. Groundwater is especially important in agricultural land uses. Water moves through the ground and carries substances from the farm into the river.

Many water bodies play an important role in the towns and cities they run through. This river in Holland is used for both recreation and industrial transportation.

LIVING TOGETHER

1.8 Read

What are Some Sources of Pollution in a River?

pollution:
substances added to air, water, or soil that cause harm to the environment.

You built and ran models of land use using a stream table. You observed the effects of the different kinds of land use on erosion, deposition, and runoff in a watershed. During the classroom discussion, you might have discovered something more. Human activities also change the quality of the water. As the water flows through the watershed, it carries with it stuff it picks up along the way. This could be dirt and soil eroded from land. Runoff and groundwater can also pick up substances like chemicals, small particles, and pieces of trash that can affect the quality of the water. Scientists call these substances that end up in the river **pollution**. Pollution can cause harm to human health or the environment. Most of the time, these substances result from human activities. Normally, they are not found in natural environments. They can be very dangerous because living organisms may not be able to handle them.

Stop and Think

Look at the pictures below and on the next page. You might have seen scenes like these in your neighborhood, or town. People are walking down the street. People are washing their cars. Some are taking care of their lawns. Someone is pouring something down a sewer drain. Workers are fixing something under the street.

Answer the following questions as you look at the pictures.

1. What kind of land use does each picture represent?

2. Identify all of the examples of human activities shown in each picture. Describe how each activity might cause pollution.

3. How might each land use affect the local runoff and groundwater?

4. Think about what might happen in each place when it rains. Describe what could be on the ground that might cause pollution in a river.

Sources of Pollution in Rivers

Once pollutants get into soil or water, they are carried by surface and groundwater to rivers. This way they are distributed over large areas, sometimes miles away from their source.

Depending on how the pollution enters a body of water, pollution sources are divided into two groups, **point-source pollution** and **non-point-source pollution**.

point-source pollution: pollution that originates from a single point or location.

non-point-source pollution: pollution that comes from many sources over a large area.

Point Sources of Pollution

Point-source pollution comes from a specific point or location. From this location, the pollution is discharged directly into rivers, lakes, or oceans. Scientists can easily identify the source of this type of pollution. They analyze the water at different points in the river or lake. The closer to the point source they measure, the higher the amount of pollutant they find.

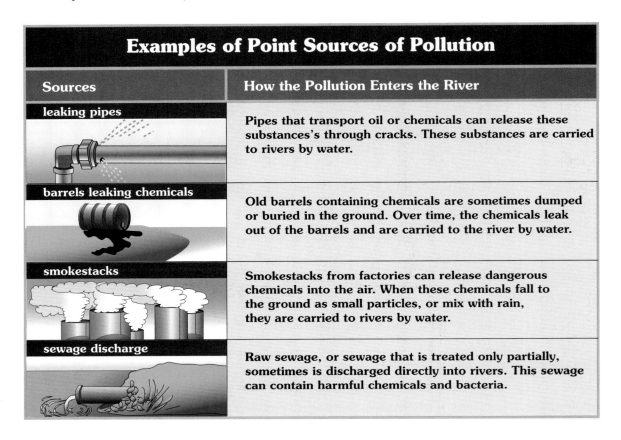

Examples of Point Sources of Pollution

Sources	How the Pollution Enters the River
leaking pipes	Pipes that transport oil or chemicals can release these substances's through cracks. These substances are carried to rivers by water.
barrels leaking chemicals	Old barrels containing chemicals are sometimes dumped or buried in the ground. Over time, the chemicals leak out of the barrels and are carried to the river by water.
smokestacks	Smokestacks from factories can release dangerous chemicals into the air. When these chemicals fall to the ground as small particles, or mix with rain, they are carried to rivers by water.
sewage discharge	Raw sewage, or sewage that is treated only partially, sometimes is discharged directly into rivers. This sewage can contain harmful chemicals and bacteria.

Non-Point Sources of Pollution

Non-point-source pollution comes from many sources and locations. Scientists cannot easily identify all the sources of this pollution. For example, one non-point source of pollution is runoff containing fertilizer used on lawns or farmland. Because the runoff has material from so many different farms or lawns, it would be difficult to pinpoint the source of the fertilizer. Another non-point source of pollution is urban runoff from roads and parking lots. Water running off these surfaces can carry oil leaked from cars or salt used to melt ice to a river. This type of pollution is often carried to the river by runoff over large areas. Non-point sources of pollution are much more difficult to control. It is hard to determine who or what is responsible for this pollution. Non-point sources of pollution can originate from a very large land area such as an entire watershed.

Examples of Non-Point Sources of Pollution	
Sources	**How the Pollution Enters the River**
fertilizer	Many people fertilize crops, lawns, and other plantings. Eventually, the fertilizer can be carried to rivers in runoff.
urban runoff	Vehicles and other equipment can leak lubricants and fuel that eventually wash into rivers.
litter	People drop or dump trash and litter in public spaces. This litter is eventually carried to waters by wind and runoff.
salt and sand	Communities spread salt and sand to prevent roads from icing over. This eventually is carried to rivers by runoff.

Reflect

Look back at the photographs you reviewed early in the Unit. The photographs show scenes of the land use you were assigned and that you modeled with your stream table. Discuss with your group the types of pollution that may result from your land use.

- Record all of the pollution sources your group identifies in the photos, including both point sources and non-point sources.

- Record any pollution sources you think might be there because of certain activities or events shown in the photos.

- For each pollution example you record, determine if it is a point source or non-point source of pollution. Make sure you write the reason why you think so.

Your teacher will lead a class discussion where each group shares their photos and their work. Listen carefully as other members of your group discuss their observations and conclusions. With your class, review and discuss the observations and conclusions drawn by other groups investigating a different land use. How are the pollution sources similar or different for each land use? Come up with a list of types of pollution you agree upon. List the types of pollution you found for each of the land uses your class has investigated.

Update the *Project Board*

The questions you focused on in this *Learning Set* were *How does water affect the land as it moves through the community?* and *How does land use affect water at it moves through a community?*. Return to the *Project Board* to update any questions or ideas you have posted. You now have many items to post in the *What are we learning?* column. Be sure to cite (in the *What is our evidence?* column) the evidence you collected to support what you say you have learned about pollution and land use. Discuss with your class what you learned and recorded on the *Project Board* that can help you answer the two questions. You might make up and discuss several new ideas or understandings that should be recorded on the *Project Board*.

What's the Point?

There are many different ways that land use can add pollutants to a watershed. All of the different ways of polluting are grouped into two different types: point sources and non-point sources of pollution. Point sources of pollution, the kind that happen in many industrial areas, are very harmful. However, they are a lot easier to stop than non-point sources of pollution. Non-point sources of pollution can be more difficult to find. They do not come from a specific place. Agricultural areas and residential areas create a lot of non-point sources of pollution through fertilizers.

Learning Set 1

Back to the Big Question and the Challenge

How Does Water Quality Affect the Ecology of a Community?

The questions you were investigating in this *Learning Set were How does water affect the land as it moves through the community?* and *How does land use affect water at it moves through a community?*. The answers to these questions might have seemed obvious at first. Now, though, you see how complicated the system of watersheds can be. You have seen how the quality of water depends on many factors. You have seen how watersheds are linked to each other.

Watersheds are an important part of understanding how water flows. Soil and chemicals might get into the runoff and be carried to the river. You have investigated how watersheds work. You also saw how runoff moves in a watershed. You noticed that runoff always moves from higher to lower elevation. Sometimes the water pools in low areas. Other times the water continues to move through the watershed to the river.

You investigated several different types of land use. You used a stream-table model to model the land cover that could be found with each land use. The stream tables helped you see erosion, deposition, and runoff for each land cover. You noticed that the water picked up some soil and carried it to the river.

In the final section, you learned some new words to describe what you saw in your model. Your models provided the beginning of your understanding. Then you read about the different land uses. Some of the issues in the land uses were similar to each other. Other issues, like those in the industrial and the agricultural uses, were very different. In all the different land uses, runoff and erosion were very important processes. Also, each land use can introduce point-source and non-point-source pollution. You identified pollution sources for each land use. You are beginning to see the connection between land use and water quality. This connection will be very important as you consider what effect the changes in Wamego might have on Crystal River.

Explain and Recommend

The *Big Question* for this Unit is *How does water quality affect the ecology of a community?* Recall that you will be answering smaller questions as you move through this Unit. Doing this will help you eventually answer the big question and address the challenge. At this point, you are not ready to answer the big question completely and thoroughly and address the challenge. But you probably have some ideas of how watershed structure, erosion and runoff, land use, and pollution sources are a part of a good answer to the big question and the challenge.

Discuss with your group what you learned and recorded on the *Project Board* that can help you answer the *Big Question* and address the *Big Challenge*. Identify ansers to the *Big Question* and recommendations you might make to Wamego town council. Using a *Create Your Explanation* page, state each of your answers and recommendations as a claim. For each claim, record the evidence you have and the science knowledge you have learned. Then write on explanation that connects the evidence and your science knowledge to your claim.

Communicate

Share Your Explanations and Recommendations

Share your claims, recommendations, and explanations with the class. If your classmates disagree with any of your claims or recommendations, discuss the evidence you have and the science knowledge that support them. Try to come to agreement. You can help one another revise your claims or recommendations to better match your evidence and science knowledge.

Update the *Project Board*

Add new claims and the evidence that goes with them to the *What are we learning?* and *What is our evidence?* columns. Add recommendations the class agrees on and the evidence that supports them to the *What does it mean for the challenge or question?* and the *What is our evidence?* columns.

Learning Set 2

How Do You Determine the Quality of Water in a Community?

You have investigated watersheds. You learned how they connect water in small areas. You also saw how those smaller watersheds are nested in larger ones. You then reviewed some water samples from a watershed. There were differences among these samples. This indicates that the water is different at different locations in a watershed.

The water from this mountain reservoir makes its way to local residents' drinking glasses.

Even though the samples were different, you know the water in the watershed is all connected. Eventually, water from separate areas will mix together. It will all flow into one larger waterway. Thus, all of the samples you reviewed earlier might mix together farther downstream.

Suppose this mixture flowed through your community, and it was your only water source. Think about how you would then judge the quality of this water. Your class began to discuss how land use in the watershed could affect what ends up in the water. Also, what is in the water can affect the plants, wildlife, and people in the community. A community should want to know exactly what is in the water. In this *Learning Set,* you will work to answer the question *How do you determine the quality of water in a community?*.

2.1 Understand the Question

Water-Quality Indicators

microbe: an organism that cannot be seen with the unaided human eye. You need to use a microscope to see it.

You may have heard about unsafe water in a community. Sometimes a community may close beaches because it is not safe to swim in the water. The water may contain dangerous **microbes** (or microorganisms). Residents may be asked to boil water before drinking it. In other cases, people may be told not to fish in certain waters. Communities must decide when the water in their area is unsafe.

Get Started

Below are some students' ideas about how a community might decide if water is safe. Read and think about each student's ideas.

Rahim: "You can just look at water and know whether you should swim in it or drink it. If it is clear, or mostly clear, it's probably fine."

Sarah: "Water that is not high quality probably smells bad. I think scientists use smell to help determine if water is of high quality."

Lucia: "Scientists test water with chemicals, like chlorine, to see what is in the water and make it clean. I think I have heard of scientists testing for acid in water."

John: "I don't really know . . . but if you look at an area of a watershed and see a lot of dead plants and animals, and people are sick, the water is probably low-quality water."

Do you agree or disagree with each student? What might you say to each student? What might you investigate to determine if a student's idea is correct? Discuss your answers to these questions with your class.

Update the *Project Board*

Your class started a
Project Board to help
you keep track of your
understanding and
questions about water
quality and ecology.
At the end of *Learning
Set 1*, you updated
the *Project Board* with
information you learned
about watersheds and
water flowing through
watersheds.

Your class has shared
their ideas about what
you need to do to
decide if water is safe.

You have also come up with ideas about investigations you could do. Now it
is time to update the *Project Board* again. You may have new questions or
new ideas about how to determine the quality of water. Record these ideas,
predictions, and explanations in the *What do we think we know?* column.
You may have some information about water-quality testing that you have
learned before or someone has told you about. Suggest it for the *Project
Board*. As you move along in your discussion, you may find that you have
disagreements about some things you think you know about water-quality
tests. Suggest things you need to learn more about. These questions could
be turned into investigations you could perform to better understand water-
quality testing. Record questions and ideas for investigation in the *What do
we need to investigate?* column.

What's the Point?

Your class might have a lot of ideas and questions about how water quality is
determined. Some of you might have some experience or knowledge in this
area. Your initial conversations and ideas helped you to become aware of
your understanding of the topic. You were able to share ideas and questions
about water quality during the *Project Board* session. Once these different
ideas are out in the open, your class can pursue investigations that focus on
these ideas and questions.

2.2 Investigate

Plant Growth as an Indicator of Water Quality

Your class has been asked to think about what is a good indicator of high water quality. You may think that the way the water looks is an important indicator. You may also have considered other indicators. Some of you may have discussed the use of chemical tests to find the quality of water. In the next few sections, your class will investigate some useful water-quality indicators. These indicators can help communities determine the quality of water in their watershed.

Fertilizers used in agriculture can affect plant growth in nearby bodies of water.

People grow plants as food. They may also grow plants to make yards and parks look attractive. Fertilizers are often used to help plants grow. These fertilizers can be carried away in the runoff from fields. The fertilizer then ends up in the watershed. There it can affect plant growth in the water. In this section, you will design an experiment that shows the effects of fertilizer on plant growth.

Design Your Experiment

Your class will work together to design an experiment about the effects of different amounts of fertilizer on the growth of a water plant, duckweed. You will make observations and record data about how the plants grow. Your observations will take place over five to ten days. You will then draw some conclusions by examining your results. Your experiment should answer the following question:

• How does fertilizer concentration affect the growth of duckweed?

Getting Started

Remember that when you investigate a process or a phenomenon, you want to learn about what factors influence its outcome. These factors are called variables. Scientists design experiments to find out how changing the value of one variable (called the *independent* or *manipulated* variable) affects the values of outcome variables (called *dependent* or *responding* variables). You have been asked to design an experiment to study the effect of fertilizer **concentration** on plant growth. Concentration, in this case, refers to the amount fertilizer you will use. You will therefore need to design your experiment so that you will be able to see how changing the amount of fertilizer (your *manipulated* or *independent* variable) will affect the growth of duckweed (the *responding* or *dependent* variable).

You will use the materials shown in the list in your experiment. Duckweed is a plant that grows in water rather than in soil. Like other plants, duckweed requires sunlight and nutrients to grow. The plant floats in the water. It gets its nutrients from the water. They get into the plant through its roots. When conditions are right, duckweed will grown quickly over a five to ten day period.

The fertilizer you will use is in liquid form. You can add fertilizer to a container of water and place duckweed in the container. The more fertilizer you add to the container, the higher the **concentration** of fertilizer. The less fertilizer in the container, the lower the concentration of fertilizer.

You will need to find a way to measure the growth of the duckweed. The most common method of measuring growth of duckweed is to count fronds. Fronds are the small leaf-like structures of duckweed.

Materials
- **4 beakers or jars**
- **4 duckweed plants**
- **water**
- **liquid fertilizer**

concentration: the amount of a substance mixed with another substance.

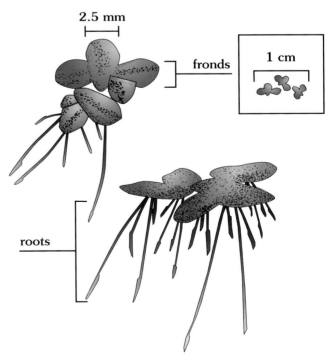

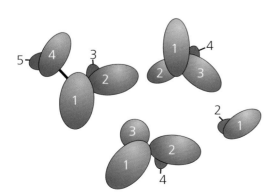

LIVING TOGETHER

When counting fronds, count *every* visible frond. Include even the tips of small new fronds. The diagram shows an example with several fronds in different positions and stages of growth. It also shows two plants joined by what looks like a stem.

You may also want to consider recording color of fronds, size of fronds, number of living plants, number of dead or dying plants, length of roots, or anything else your class decides is important. Draw sketches of what you observe.

Planning Your Experiment

To get started, each group will plan an experiment to answer the question *How does fertilizer concentration affect the growth of duckweed?*. Then, after examining the different experiment ideas, the class will agree on one experiment to run. Remember to discuss and record answers to the following questions as you are planning your experiment.

Question

- What question are you investigating and answering with this experiment?

Prediction

- What do you think the answer is, and why do you think that?

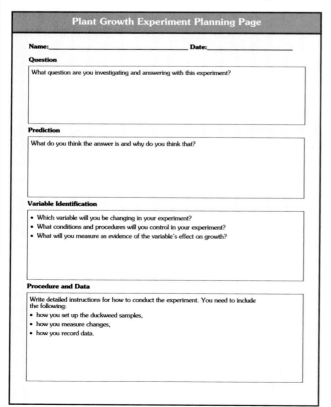

Variable Identification

- Which variable will you be changing in your experiment?

- What conditions and procedures will you control in your experiment? That is, what conditions and procedures will you keep the same as you change your variable?

- What will you measure as evidence of the variable's effect on growth?

Procedure and Data Collection

Write detailed instructions for how to conduct the experiment. You need to include

- how you set up the duckweed samples,

- how you measure changes, and

- how you record data.

Also, you should explain how you determine whether or not you can trust the data.

You will be given a *Plant Growth Experiment Planning* page. The series of questions on this page will help you plan. With your group, discuss and fill out the sections of this page. Use the hints on the planning page as a guide. Be sure to write enough in each section so that you will be able to present your experiment design to the class. The class will want to know that you've thought through all of the parts of your plan.

Communicate Your Plan

Plan Briefing

To help you as you learn to design experiments, you will share your investigation plan with the class during a *Plan Briefing*. As you are finishing your design plan, begin to draw a poster for presentation of your design plan to the class. Your teacher will provide you with a large sheet to use for your *Plan Briefing* poster and possibly a template to follow.

Others in the class have planned experiments to answer the same question you are answering. Discuss your proposed experiments with the class. Be very specific about your design plan and what evidence helped you make your design decisions. You will probably see differences and similarities in these plans. This discussion will allow you to compare ideas. You will avoid making mistakes that others can see in your plan. You will talk about what is good about each plan. You should also discuss what needs to be improved. Be very specific about your design plan and what evidence helped you make your design decisions.

After you dicuss all the plans, the class will decide together how to set up the experiment and what to measure. You will decide how to set up the experiment and what to measure.

Run Your Experiment

Once you have decided on a class experiment, it will be time to set it up and run it. Set it up as your class has decided. Be sure to place the jars in an area where the duckweed can get the light it needs to grow. You will observe the jars and collect data every day for the next six to ten days.

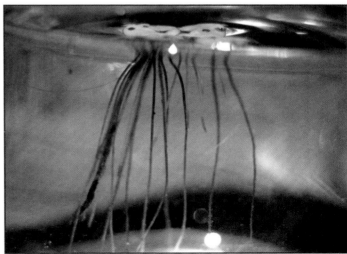

How does duckweed respond to fertilizer added to the water?

Recording Your Data

You will be recording data about the duckweed's growth. Record your data on a *Plant-Growth Data and Observations* page. There is space on it for recording the number of fronds, their color, the number of dead or dying plants, and anything else your class decided is important to observe and record. There is also space for you to sketch what you observe.

> ### Plant-Growth Data and Observations
>
> Name: _____   Class: _____
>
> Use the following pages to record the plant growth in the four jars over a period of 5-10 days.
>
> Observations may include the number of fronds, color of fronds, size of fronds, number of living plants, number of dead or dying plants, length of roots, or anything else your class decides is important.
>
> Draw sketches.
>
> Date: _____
>
> Observations

● **You will need five to ten days to gather the data from this experiment. During that time, you will look at two other indicators of water quality in the next sections. Then you will return to analyze the data you have collected here.**

Under certain conditions, duckweed will grow until it cover's a pond's surface.

Project-Based Inquiry Science

Fertilizers Contain Nitrates and Phosphates

Fertilizer is high in chemicals called nitrates and phosphates. A small concentration of these chemicals is good for the health of a river, lake, or stream. Plants need nitrates and phosphates to grow. However, high levels of nitrates and phosphates in the water can cause problems.

Algae and other plants that grow in the water use phosphates. However, they need only a small amount of phosphate to grow. Too much phosphate can cause large amounts of these plants to grow. Large growths of algae are called algal blooms. Algal blooms can also result from too many nitrates in the water. Algal blooms can cause problems for many reasons. They can keep light from getting to the roots of plants in the water. Also, when large growths of algae die, bacteria break them down. This makes the oxygen levels in the water decrease. In turn, fish and other animals might die.

Fertilizers and some laundry detergents are major sources of phosphates and nitrates. Excess nitrates, especially near cities or farms, can also come from sewage.

The table shows the phosphates and nitrates that can be found in water of different qualities. The "ppm" stands for *parts per million*. This term is used when only extremely small amounts of the substance are found in another substance. It means that one particle of the substance is found for every 999,999 other particles. That's like adding a drop of food coloring to 150 L (40 gallons) of water. But, as you can see from the table, even 3 or 4 parts per million of phosphate is bad for water quality. This table can help you determine the quality of any samples of water that you may want to investigate.

algae (singular, alga): simple organisms that live in water. Some can be as small as one cell. Some are made up of many cells. Algae made of many cells may be called "seaweed."

Water Quality and Concentration of Phosphates and Nitrates		
Water Quality	**Phosphate**	**Nitrate**
excellent	1 ppm	0 ppm
good	2 ppm	0 ppm
fair	3 ppm	1–20 ppm
poor	4 ppm	30 ppm or more

"ppm" measures concentration in parts per million.

Stop and Think

1. What are three sources of phosphates and nitrates in water?

2. Why are too many nitrates and phosphates unhealthy for a body of water?

Analyze Your Data

To help you analyze the data from this investigation, answer the following questions. Be prepared to discuss your answers with the class.

1. What effects did the fertilizer concentration have on the growth of the plants? Support your answer with evidence from the investigation.

2. Describe the relationship between fertilizer concentration and duckweed growth. (Summarize the trends you found in each of the different solutions.)

What's the Point?

Your class has been recording data about the growth of duckweed. You set up an investigation to see how fertilizer concentration affected the growth of duckweed (a water plant). Fertilizers contain nitrates and phosphates. Nitrates and phosphates are needed for plant growth.

However, when large quantities of these chemicals enter the water through runoff or groundwater, they can cause problems. They increase the growth of plants that live in the water. This can lead to the death of fish and other animals. It may even lead to the death of the plants themselves. Plant growth can be used as an indicator of the amount of nitrates and phosphates in water. If you see a thick plant growth in water, you know that water contains large amounts of nitrates and phosphates.

The amount of fertilizer in the water also can have a large impact on dissolved oxygen. Bacteria break down plants when they die. This results in reduced oxygen levels in the water. In turn, the fish and other animals that live in the water may die because there is not enough oxygen for them to survive.

2.3 Investigate

pH as an Indicator of Water Quality

In this *Learning Set*, you have been thinking about the quality of water. You were asked to consider what is a good indicator of water quality. You may have decided that the way water looks is an important indicator. However, some pollutants are invisible. You also have begun looking at indicators of water quality that allow you to identify pollutants that are too small to see. In this section, you will investigate **pH** as an indicator. pH measures whether a substance is an **acid** or not. When substances have a pH that is either very high or very low, these substances can be harmful to many living things, including you.

pH: a measure of how acidic a substance is.

acid: a solution with a pH of less than 7.

pH indicator: a chemical that can be added to a solution to determine pH.

Demonstration

At the beginning of this Unit, you looked at five different samples of water. You may have at first thought that the water in Jars 3 and 4 was of a high quality. Perhaps you thought that you could drink the water in these jars.

Your teacher will add a special chemical to the water in Jars 3 and 4. This substance is called a **pH indicator**. It will tell you something about the quality of water. This substance will also be added to a sample of water in a third jar. The water in this jar is distilled water. You can think of it as "pure" water.

Observe

Answer the following questions while observing the demonstration.
Be prepared to discuss your answers with your group and the class.

1. When looking at the three jars of water that your teacher used in the demonstration, could you tell them apart? Why or why not?

2. After the indicator was added to each jar, was there a difference in the water in each jar? Describe what you saw.

3. Here is a picture of water from a clean swimming pool.

   a) How does the water look? Compare it to the water in the jars in the demonstration.

   b) Would you swim in this water? Explain your answer.

   c) Would you drink this water? Think about the way water in a pool smells when you explain your answer.

pH scale:
a scale used by scientists to measure the acidity of a solution.

neutral:
a solution with a pH of 7.

Acids and pH

Acids

You have probably heard about acids. You might have seen acids in bottles that have poison or caution labels on them. Acids may seem dangerous. However, many acids are very important to life. For example, the liquid in your stomach is acidic. It makes it possible for you to digest food. Vitamin C is acetic acid. It is important to maintaining a healthy body.

Too much acid can have negative effects. Sometimes your stomach makes more acid than it needs to digest food. This can cause indigestion. Outside your body, acids react with metals and cause them to corrode. Some plants need acidic soil. However, too much acid in water can have negative effects on the plant and animal life in an aquatic ecosystem. In the atmosphere, acids react with other chemicals to create acidic water. This water can then fall as "acid rain." Acid rain causes environmental damage by changing the pH of bodies of water, corroding metals, and mixing with substances it contacts to create other pollutants.

The pH scale

Scientists use the **pH scale** to measure how acidic a solution is. The scale goes from 0 to 14. Solutions with a pH less than 0 are acidic.

Solutions with a pH of 7 are **neutral**. Solutions with a pH greater than 7 are called bases. (You will read about them later.) Battery acid is very acidic. It has a pH of 0. Lemon juice is very acidic but not as acidic as battery acid. It has a pH of about 2. Orange juice and tomatoes are very acidic but not as acidic as lemon juice. They have a pH around 4. Milk is not very acidic. Its pH is between 6 and 7. Distilled (pure) water has a pH of 7.

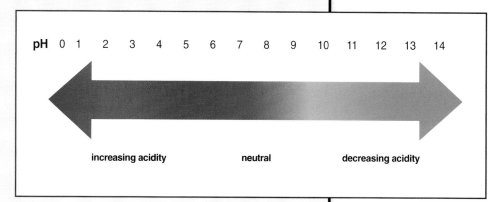

| pH | 0 | 1 | 2 | 3 | 4 | 5 | 6 | 7 | 8 | 9 | 10 | 11 | 12 | 13 | 14 |

increasing acidity　　　　neutral　　　　decreasing acidity

PH measures the number of hydrogen ions in a liquid. Hydrogen ions are very small particles. When something is more acidic, it has more of these particles. Notice that low numbers of pH actually indicate more hydrogen ions. The lower the number, the more acidic a solution is. Higher numbers indicate lower concentrations of hydrogen ions and less acidity. The explanation for this is very complicated. It is not necessary for you to understand it at this point. So for now, just try to remember this odd fact.

There is one more important fact about pH to remember. When the pH of a substance changes by one number on the scale, the level of acidity changes by a factor of ten. (pH stands for the power of hydrogen.) For example, a substance with a pH of 2 is *ten times* more acidic than another with a pH of 3.

pH Indicators

In the demonstration you saw, your teacher used a pH indicator. A pH indicator is a substance that is sensitive to pH. A very small amount of indicator can be added to a water sample. As the indicator and water mix, you will see a change in color. The color tells you the acidity level, or pH, of the water sample. There are many different indicators that scientists use.

Many plants contain indicators. Flowers such as hydrangeas have different color petals depending on how acidic the soil they are grown in is. Red cabbage juice can also be used as an indicator. It will change the color of a water sample. The different colors are shown in the chart on the next page. Each color corresponds to a pH number or acidity level.

Hydrangea petals are different colors depending on the acidity of the soil in which they are growing.

You will be working with very special substances and equipment. It is important that you are very careful in handling these items to insure that you and your classmates remain safe. Your teacher will provide further instructions about science safety during your investigation.

Procedure

In Part A, you and your group will investigate and determine the pH of five solutions with an indicator provided by your teacher. In Part B, you will attempt to change the pH of another sample by mixing it with a sample with a different pH. Follow the directions below to conduct your investigation. You will work as a group, but each of you will need to complete the data chart.

Part A: Identify Acidity

1. Place four drops of indicator in each of the samples in the test tubes you receive from your teacher. Very gently, swirl each test tube to mix the indicator. Place the test tube back in the rack before you make any observations. If the color is too faint, add one to two more drops of indicator.

2. Record the color of each test tube in your data chart.

3. Match the color of each sample with one of the colors in the chart provided here. Record the pH paired with the matching indicator color on a data table similar to the one shown.

pH	0–2	3–4	5–6	7–8	9–10	11–12	13–14
cabbage juice color	red	red-pink pink	pink-purple	pale blue	yellow green	pale green	green

	Sample 1	Sample 2	Sample 3	Sample 4	Sample 5
color					
pH					

Part B: Mixing Solutions Together

1. Add five to six drops of indicator to Solution A in the beaker provided by your teacher. Gently swirl the beaker. What color does the solution turn? Using the chart, determine the pH of this solution.

2. Draw a full pipette of Solution B, also provided by your teacher. Squeeze the full pipette of Solution B into Solution A. Repeat this three times. Observe what happens to the color of Solution A each time you add more of Solution B.

3. Continue to add Solution B to Solution A with the pipette until you have added 50 mL. What is the color of Solution A at this point? Use the pH color chart to determine its pH.

4. Very slowly and gently, pour the remaining 50 mL of Solution B into Solution A. What is the color of Solution A at this point? Determine the pH according to the chart on the previous page.

Analyze Your Data

1. How were the five samples you had in front of you similar to Jars 3 and 4?

2. Which of the sample or samples are most acidic?

3. Which of the sample or samples are least acidic?

4. Which is most like Jar 3 from the demonstration? What do you think is the pH of Jar 3?

5. Which is most like Jar 4 from the demonstration? What do you think is the pH of Jar 4?

6. As you added Solution B to Solution A, what was happening to the pH of Solution A? What evidence do you have of this?

7. What substances do you think you were mixing in Part B of the pH investigation?

8. How could you return the pH of Solution A to its original condition?

9. Solution A became _____ times less acidic during the investigation.

More about pH

The pH of liquid substances can be measured and reported as a number on the pH scale. The pH scale on the next page runs from 1 to 14. The diagram on the next page shows you the pH of several common substances.

At the left side of the scale you see lemon juice and vinegar. You may have tasted these substances before, and if so, you know they are very

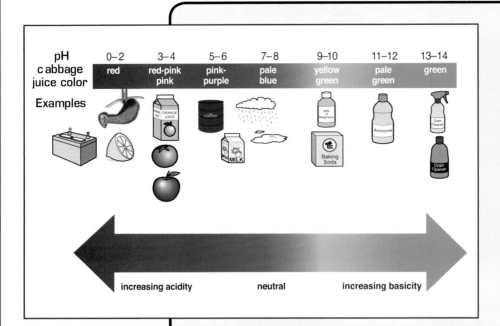

sour. These are acidic substances.

Look at the right side of the scale. **Basicity** is a term used to describe non-acidic substances. Alkaline substances are **bases**. Antacids are substances that help people with upset stomachs. The process of digestion can create a very acidic condition in the stomach. People take antacids to reduce stomach acidity. Bases feel slippery, like soap.

Extremely alkaline substances, with very high pH, are used to make cleansers. These cleansers kill off bacteria that feed on substances collected on surfaces.

Substances that have a low pH (1–6) can become less acidic when they are mixed with substances that have higher pH (8–14). Likewise, substances that have a high pH can become more acidic when they are mixed with substances that have low pH. If the mixture is just right, the substance can become neutral, with a pH of 7. This is often called a pH-balanced substance.

Some **aquatic organisms** are very sensitive to pH. (Aquatic organisms are organisms that live in water.) They may die if the pH of the water changes even slightly. Other aquatic organisms are more resilient and can tolerate a wider range of pH levels. The table shows the pH reading of different qualities of water. The chart on the next page shows the range of pH that different organisms can tolerate.

basicity: a term used to describe non-acid substances.

base: a solution with a pH greater than 7.

aquatic organisms: organisms that live in water.

Water Quality and pH	
Water Quality	**pH Reading**
excellent	7
good	6 or 7
fair/poor	less than 5 or greater than 7

Range of pH that Different Organisms Can Tolerate		
Organism		**pH** 0 2 4 6 8 10 12 14
bacteria		
plants and algae		
catfish, carp, and some insects		
bass and bluegill		
snails and clams		
many fish and insects		
trout		

What's the Point?

You now know that pH is the level of acidity of a substance or solution. The pH of water is very important. Living things can only survive within certain ranges of pH. When the water changes pH, especially when it happens suddenly, it can kill the living things in and around the water.

The pH of water can change when the water is contaminated by industrial waste, agricultural and urban run-off, or drainage from mining operations. The type of contamination depends on the way the land is used.

Why do you need to know the pH of the water? The pH can tell you whether there might be a source of acid or basicity in your stream. Most organisms prefer water with mid-range pHs (6.0–7.5). Water within this pH range is considered of high quality and is close to neutral. Trout, which are important to Wamego and St. George, can only live in a very narrow range of pH.

2.4 Investigate

Dissolved Oxygen as an Indicator of Water Quality

O_2

So far you have looked at plant growth and pH as indicators of water quality. Dissolved oxygen (DO) is another important indicator of water quality. All organisms that live in the water need oxygen to survive. Testing for DO can help you determine the health of a body of water.

The oxygen molecule is two oxygen atoms bonded together. The formula for an oxygen molecule is O_2 (O-two). Dissolved oxygen is oxygen molecules surrounded by many water molecules. You cannot see these molecules because they are too small. However, aquatic animals use DO for respiration (breathing).

Most of the DO in water comes from the atmosphere. As water moves along, oxygen is mixed into it. Another source of DO is aquatic plants. Oxygen is formed by plants in the process of photosynthesis. Excellent water has high levels of dissolved oxygen. Natural changes can affect the amount of DO. Humans can also affect the "health" of the water.

Demonstration

Your teacher is going to demonstrate how to measure the amount of DO in several water samples. You will investigate the effect of two different factors, temperature and **turbulence**, on the amount of dissolved oxygen.

Most of the dissolved oxygen in water bodies comes from the atmosphere, but some comes from plants in the water.

Dissolved oxygen is best measured in a unit called *percent saturation*. This indicates how much dissolved oxygen is in a sample compared to what it can hold. For example, water at 28°C (82°F) can hold 8 ppm of dissolved oxygen. When it has 8 ppm of oxygen in it, it is 100% saturated. If that same sample of water had only 4 ppm of dissolved oxygen, it would be 50% saturated.

Temperature and Dissolved Oxygen

Your teacher will set up four samples of water at different temperatures and use a temperature probe to measure the temperature of each sample. An oxygen probe will be used to measure the amount of oxygen in each sample.

Make your observations. Record them in a table similar to the one shown.

Dissolved Oxygen and Temperature				
	Sample 1	**Sample 2**	**Sample 3**	**Sample 4**
temperature				
dissolved oxygen				

Turbulence and Dissolved Oxygen

Your teacher will use the same samples as in the previous demonstration. The dissolved oxygen in each sample will be measured before and after stirring.

Predict what will happen to the dissolved oxygen when the samples are stirred.

Make your observations. Record them in a table similar to the one shown.

turbulence: the violent disruption, agitation, or stirring up of something (in this case, of water).

Dissolved Oxygen and Stirring					
		Sample 1	**Sample 2**	**Sample 3**	**Sample 4**
temperature					
dissolved oxygen	before stirring				
	after stirring				

Analyze Your Data

1. Describe the trends you see in these results.

2. What appears to be the relationship between temperature and dissolved oxygen found in water?

3. What appears to be the relationship between turbulence (stirring) and dissolved oxygen found in water?

4. When would a river be stirred up like the water in the jar?

Dissolved Oxygen

Temperature, turbulence, and amount of plant growth all affect the amount of oxygen dissolved in water.

Temperature

Colder water can hold more dissolved oxygen than warmer water. In fact, water at 8°C (46°F) can hold 12 ppm dissolved oxygen. Compare this to only 8 ppm in 28°C (82°F) water. You should have seen a clear trend in your results suggesting this relationship.

Turbulence

The more turbulent the water is, the higher the amount of dissolved oxygen is in it. As water is mixed and splashed, it traps oxygen from the air. This oxygen dissolves into the water. Water that is moving rapidly and mixes with air has more DO than water that is still.

Rivers should be saturated with dissolved oxygen. That is because rivers are always flowing. Rivers that fall below 90% saturation are considered to be less than desirable. The table on the next page shows the amount of DO in different qualities of water.

Plant Growth

Large numbers of plants in and around a body of water can rob water of its dissolved oxygen. When plants and other organic material (anything that is or was once living) falls into water, the material begins to break down (decompose). Bacteria decompose the material. In the process of doing this, bacteria use up dissolved oxygen. Thus, a water body with a large amount of plant growth around it will eventually have very low levels of dissolved oxygen.

Water Quality and Dissolved Oxygen	
Water Quality	**Dissolved Oxygen (% of saturation)**
excellent	91 – 100%
good	71 – 90%
fair	51 – 70%
poor	50% or less

What's the Point?

The amount of oxygen (O_2) that can dissolve into water depends on two factors. One is the temperature of the water. Cool waters can dissolve more oxygen than warm waters. The amount of dissolved oxygen (DO) also depends on turbulence. Quickly flowing, turbulent areas, like waterfalls, have high levels of DO. By comparison, slowly moving or still water, like lakes and ponds, will have lower levels of DO.

Aquatic organisms need DO to survive. Fish use their gills to obtain DO from the water. They need oxygen to breathe and live. The organisms that live in a body of water depend on the amount of DO in the water to live. Generally, the more DO, the greater the number of organisms that can live in the water.

Large amounts of plant growth in water or very near water can rob the water of its DO. This is because when the plants die, bacteria use DO as they decompose the dead plants.

2.5 Read

More Water-Quality Indicators

You have investigated nitrates and phosphates, pH, and dissolved oxygen as indicators of water quality. Now you will read about three more indicators of water quality. You will read about and discuss the indicators temperature, turbidity, and fecal coliform.

Temperature

Temperature is very important to water quality. As you saw in the previous section, temperature affects the amount of DO in the water. It also affects the rate of photosynthesis. Under certain conditions, photosynthesis increases at higher temperatures. Therefore, the amount of plant growth increases. This can be a good thing. Photosynthesis adds DO to the water. However, plant growth can increase too much. The plants will eventually die. Then, decomposing bacteria will use up the DO.

Science Connection

Thermoregulation is the ability of an organism to maintain a fairly steady body temperature. Some organisms do this by living in environments with appropriate temperatures. Other organisms rely on their bodies to produce or dispel heat.

These carp need water temperatures of 15° – 34°C (59° – 93°F) to live.

Temperature also affects the balance of an ecosystem. Most organisms are suited to live in a given range of temperature. Some organisms, such as trout, prefer cooler water. Others, such as carp, need warmer conditions. As the temperature of a river or lake increases, cool-water animals will leave or die. Warm-water animals may replace them. Most organisms cannot survive in temperatures of extreme heat or cold. Temperatures that are too high or too low can stress an organism. This stress makes organisms more prone to disease. It also makes it more difficult for them to react to pollution.

thermal pollution: a change in temperature in a natural body of water that is caused by humans.

turbidity: how cloudy, murky, or opaque something is.

A Secchi disk can be used to measure turbidity. The disk is lowered and raised in the water. A depth reading is taken at the point where the disk disappears and reappears.

Humans cause many changes in the temperature of a body of water. This is called **thermal pollution**. One source is warm water added by industries. Scientists test temperature by measuring the temperature of the river, lake, or stream at its source. Scientists then move to another point in the river or lake. They take a second measure of the temperature. If the difference in temperatures is very large, there is a problem. The table shows the water quality for various changes in temperature.

Water Quality and Temperature	
Water Quality	**Temperature Change Between Samples**
excellent	0–2°C (0–4°F)
good	3–5°C (5–9°F)
fair	6–10°C (10–18°F)
poor	10°C (18°F) or greater

Turbidity

How clear water looks can also help you determine the quality of the water. **Turbidity** is the measure of how clear water is. Materials such as clay, silt, organic and inorganic matter, and microscopic organisms can be suspended in water. These cause turbid, or murky, water. The murkier the water, the greater the turbidity.

Not as many types of organisms can live in turbid water as can live in clear water. This is because there is less sunlight that can get through the water. Also, the suspended particles absorb sunlight. Therefore, water temperature increases. This causes oxygen levels to fall. (Remember, warm water holds less oxygen than colder water.)

Turbid water may be the result of soil erosion, urban runoff, and/or bottom sediment disturbances. Bottom sediment disturbances can be caused by boat traffic. They can also result from large numbers of bottom feeders. (Bottom feeders are organisms that stay near the bottom of the water when feeding.)

Fecal Coliform

Fecal coliforms are bacteria. They occur naturally in animals' digestive tracts. They aid in the digestion of food. Fecal coliform bacteria are found in feces. They appear in the feces of humans and other warm-blooded animals such as cattle and birds. Sometimes there is too much of this type of bacteria present. Then there is the possibility that harmful microbes are also present. These microscopic organisms or viruses can cause disease.

Even though there may be very few harmful microbes, it only takes a small amount to make a person sick. In water, if fecal coliform counts are high, there is a greater chance of harmful microbes being present. Swimming in waters that have high fecal coliform counts can increase a person's risk of getting sick. Microbes can enter the body through the skin, cuts, nose, ears, or mouth. Fecal coliform bacteria can enter a river through runoff or sewage discharge.

There is a very simple determination of water quality when it comes to fecal coliform. Water that tests negative (no fecal coliform) is good quality. Water that tests positive for fecal coliform is considered poor quality. It is that simple.

Sewage discharge that is not treated in some way can introduce harmful bacteria into waterways.

Stop and Think

1. In the previous section, you observed how temperature affects the amount of dissolved oxygen. What trend did you observe?

2. How can increased plant growth both increase and decrease dissolved oxygen in water?

3. An ecosystem includes the organisms that live in it. How can changes in temperature in an ecosystem change the organisms that live there?

4. What is thermal pollution?

5. How do scientists measure the quality of water using temperature?

6. Why is it important to measure the turbidity of the water?

7. What causes turbid water?

8. Why is fecal coliform measured to determine the health of a river?

9. How do fecal coliform bacteria end up in a river?

10. What role do humans have in altering temperature, turbidity, and amount of fecal coliform in water?

What's the Point?

These three indicators, temperature, turbidity, and fecal coliform, are critical to understanding the quality of a body of water. Each can easily be affected by the different land uses you and your classmates have been learning about.

Now that you have reviewed six indicators, you understand much better how scientists can evaluate the quality of water in a community. Next you will be applying your knowledge of these indicators to diagnose possible water-quality problems in a watershed.

Learning Set 2

Back to the Big Question and Challenge

How does Water Quality Affect the Ecology of a Community?

Over the past few days, your class has examined several water-quality indicators:

- pH
- dissolved oxygen
- nitrates and phosphates
- temperature
- turbidity
- fecal coliform

You also learned the factors that can lead to changes in these indicators.

During previous activities, you looked at how land use can affect the waters in a watershed. Your group has spent some time looking at one of four land uses: residential, commercial, industrial, and agricultural. As a wrap-up to this *Learning Set*, you will apply what you have learned about indicators to your assigned land use.

In the beginning of this *Learning Set* you were asked the question *How do you determine the quality of water in a community?.*

You will now answer this question, but not in the way you might expect. During this activity, your group will determine which water-quality tests would be best for the land use you were assigned in *Learning Set 1*. You will review the photos of your land use. Your group will discuss and determine the water-quality tests you should conduct. You will base your decisions on features of land use found in the photos.

Each test you decide to conduct needs to have an explanation justifying the use of the test. You will need to

- use evidence from the photos to suggest why the test is appropriate.

- describe what issue or problem your test is intended to identify.

For example, suppose that one of your pictures features a car wash that dumps used water into a creek that runs next to the car wash. Your group might decide that the detergent in the car wash water would do something to disrupt the water quality of the creek, specifically pH. You have several pieces of information and some understandings learned from your pH investigation and research.

- You know from your investigation of pH that the water should not have a pH outside the 6.0–7.5 range.

- Your group also read that detergents can have a pH around 10–11.

- Your investigation showed how the pH of a solution can change as it mixes with a different solution of lower or higher pH.

- Your research shows that many living organisms cannot survive changes in pH easily.

Thus, when the soapy water mixes with the creek, the creek's pH could move above desired levels. So, your group might decide to do a pH test to see if the pH of the creek is a problem.

Your teacher will provide you with an *Applying Indicators to My Land Use* page. Review how the example above would be recorded on this sheet.

Applying Indicators to My Land Use

What water quality indicator do you want to test?

- *pH*

What feature or aspect in the photos suggests you should test this?

- *The car wash is dumping soapy water into the creek.*

What do you predict would be the outcome of the test? Provide a specific number or value you expect to see from the test.

- *The pH might be as high as 8 or 9.*

What facts or useful information do you have from your research and investigations that suggest this is a good test and an important one for good water quality? Provide at least two.

- *Water should not have a pH outside the 6–7.5 range.*
- *pH of a solution can change as it mixes with a different solution of lower or higher pH.*
- *Many living organisms cannot survive quick changes in pH.*

What process or condition that occurs in the land use could cause possible problems with this indicator?

- *Detergents can have a pH around 10–11, so the dumping of the soapy water might be changing the pH of the creek to unsafe levels.*

Communicate Your Ideas

Idea Briefing

An *Idea Briefing* is like a *Solution Briefing*. It is a presentation that allows presenters and audiences to communicate effectively about an idea or solution. The *Idea Briefing* provides an opportunity to share what people have tried and what they have learned from their attempts at solving a challenge.

Your *Idea Briefing* should focus on

- the factors that exist in your land-use photos that suggest there might be a problem;
- the water-quality tests you want to perform; and
- the possible problems you might expect in the indicator's test results.

Your poster should address these three points. Use the information you recorded on the *Applying Indicators to My Land Use* page to make your poster. Your teacher will give more guidance on how to lay out your poster and include important information.

Be sure to fill out an *Idea Briefing Notes* page as you listen to everyone's presentation. Record the evidence people cite for justifying the test of an indicator for each land use.

Update the *Project Board*

The question for this *Learning Set* was *How do you determine the quality of water in a community?* The *Idea Briefing* should have helped you think about the water quality tests and how they might help you determine water quality. Discuss what you've learned with your class. You might formulate and discuss several new ideas that should be recorded on the *Project Board*. Record your new understanding in the *What are we learning?* column of the *Project Board*. Be sure to include evidence from your investigations.

Now, return to the big question and the challenge. Think about the connection between what you have learned about water quality and the *Big Question*. Describe the connections between your learning and the *Big Question* in the last column of the *Project Board*.

Learning Set 3

How Can Changes in Water Quality Affect the Living Things in an Ecosystem?

ecosystem: all the living things in a given place, along with the nonliving environment.

interaction: a kind of action in which two or more organisms have an effect on each other.

ecology: the study of the relationships between organisms and their environment.

You have studied watersheds and water quality. You modeled how water can affect the land over which it flows. You also modeled how a river is affected by changes in land use. Changes in land use affect how water flows in the watershed. They also affect the quality of the water.

Factories might use water from a river. They may return it to the river at a higher temperature. Runoff from farms and lawns can increase the amount of phosphates and nitrates in the river. You learned about how this affects the quality of the water. The quality of water, in turn, can affect the organisms that live in and around it.

A watershed, the land, the water flowing through it, and the plants and animals living in it are all part of what scientists call an **ecosystem**. An ecosystem is made up of both nonliving and living components. In an ecosystem, the living things have **interactions** with one another and with the nonliving things.

What do you think this otter depends on to live?

Human activities, water quality, and the types of organisms that live in the watershed are all connected. When you understand the interactions among living and nonliving things in an ecosystem, you can better measure the effects your actions might have on the **ecology** of the community.

In the first two *Learning Sets*, you investigated the nonliving parts of an ecosystem—watersheds and how they change. In this *Learning Set*, you will look more closely at the living things in an ecosystem. You will answer the question *How can changes in water quality affect the living things in an ecosystem?*.

3.1 Understand the Question
Thinking about Ecosystems

abiotic: nonliving.

biotic: living.

habitat: the place where an organism lives and grows naturally.

community: groups of organisms living together in a certain area. The organisms interact and depend on one another for survival.

An ecosystem is made up of both nonliving and living parts. The nonliving parts are called **abiotic** components. They include things like water, soil, oxygen, temperature, light, and chemicals. The living parts are called **biotic** components. They include organisms like plants and animals.

The environment around a river can change a great deal from stream to stream. It can also change in different locations along the same river. As the environment, or **habitat**, changes, the types of organisms that live there also change. Scientists call a group of organisms living together a **community**. To understand how organisms interact with their environment and one another, scientists collect and organize lots of information about the organisms in a community and the environment in which they live.

Science Connection

A group of ecosystems that have the same climate and similar communities is called a biome. You can learn more about biomes if you turn to the end of this *Learning Set.*

Update the *Project Board*

Your class started a *Project Board* to help you keep track of your investigations and questions regarding water quality in a community. At the end of *Learning Set 1,* you updated the *Project Board* with information about water moving in a watershed. After you completed *Learning Set 2,* you added the results of your investigations on water quality.

Consider what you might like to know about the biotic parts of the aquatic ecosystem you have been talking about. What are some ideas you have about the types of organisms that live in the aquatic ecosystem? Discuss what you think you know about how living things interact in nature. Maybe you have some ideas already about how living things can be impacted

by water quality and how those impacts could affect other organisms in an ecosystem. Your teacher will help your class to discuss your ideas and questions and then record them on the *Project Board*.

What's the Point?

In this section, you thought about ecosystems. An ecosystem includes all the living and nonliving parts of an environment. You also considered what you might think you know about the types of living things in an aquatic environment. You have updated the *Project Board* to record the questions you have and would like to investigate. Your class might have a lot of ideas and questions about how water quality can affect living things. Once all the different ideas are recorded, your class can pursue investigations that focus on these ideas and questions.

3.2 Investigate

Classifying Macroinvertebrates

The presence of different kinds of animals can indicate a healthy ecosystem.

When scientists investigate the characteristics of an ecosystem, they look at both the living and nonliving things. Ecosystems that are located in a body of water are called **aquatic ecosystems**. Sometimes scientists specifically study the different types of animals that live in the water. Consider some of the animals you know that live in an aquatic ecosystem.

Fish are aquatic animals. So are frogs. There are a lot of different animals that live in the water. Many of these animals you have probably never seen. Often, scientists are interested in finding out exactly what kind of animals live in streams. They want to identify how many different types of animals live in the water. This is called **diversity**. They also want to know how many of each type of animal is in the stream. This is called **abundance**.

aquatic ecosystem: an ecosystem located in a body of water.

diversity: the different types of animals.

abundance: the number of a type of animal.

classify: arrange or sort by categories.

To understand the diversity and abundance of different animals in the water, scientists have to sample the water. They pick a spot in the water and collect samples of water with the animals in it.

Then they pick all the animals they can see out of the water. They sort them, or **classify** them, based on how they look. Next, the scientists take each group of animals and count how many of each one they have found. They count the abundance of the animals.

Procedure

1. You are going to watch a video of scientists working. They are collecting river-water samples to classify and count the animals that live in the water. As you watch the video, think about the types of animals they are trying to collect. Look at the kinds of tools they are using. Think about why they use those tools to collect the animals. When the video is over, make lists of the types of animals they were trying to collect, the tools they used, and why they used those tools.

macroinvertebrate: an organism that does not have a backbone and can be seen with the naked eye.

2. Scientists group the animals they find in the water by how they look. They put the animals that are similar to one another together in the same group.

One group of animals that live in the water is called the **macroinvertebrates**. *Macro* means that you can see it with your eyes. You don't need any tools like a microscope or hand lens to help you see it. Invertebrate means that the animal does not have a backbone. Common macroinvertebrates that live in the water include all kinds of insects, snails, and crustaceans such as crayfish.

Your teacher will give you samples of several different macroinvertebrates. These samples are similar to the ones collected by the scientists you observed in the video. Observe the differences in the animals. Look carefully at their body parts.

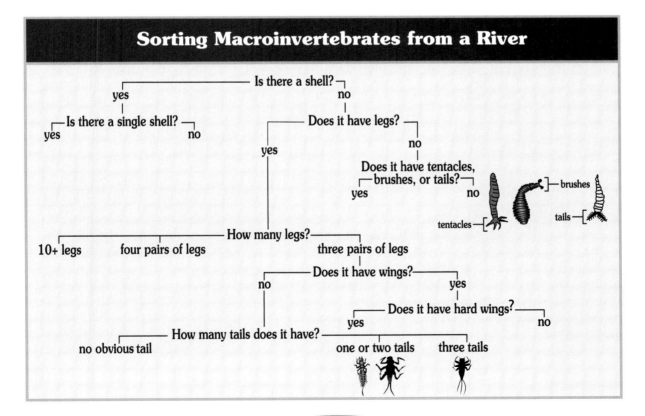

Sorting Macroinvertebrates from a River

Is there a shell?
- yes — Is there a single shell? — yes / no
- no — Does it have legs?
  - yes — How many legs?
    - 10+ legs
    - four pairs of legs
    - three pairs of legs — Does it have wings?
      - no — How many tails does it have?
        - no obvious tail
      - yes — Does it have hard wings?
        - yes — one or two tails
        - no — three tails
  - no — Does it have tentacles, brushes, or tails?
    - yes — tentacles
    - no — brushes / tails

LIVING TOGETHER

Identifying Macroinvertebrates		
Picture	**Characteristics**	**Name**

Use the questions on the key on the previous page to help you sort your macroinvertebrates. Place each macroinvertebrate under the last question that sorts it from the rest.

As you follow the chart and answer each question, record your answers in a table similar to the one above. One example has been done for you. Your teacher will lead you through another example as a class. Then your group should meet and attempt to sort each of the samples provided.

3. Scientists name living things so that they can talk to one another. They need to know that they are talking about the same animals. You have grouped your macroinvertebrates, but you have not named them.

   Your teacher will give your group another key. This key is very similar to the one in your book. However, this one also includes pictures and names of the animals. Match your animals with the pictures and names on the new key. Record the name of each macroinvertebrate in your table. Your teacher will lead you through another example as a class. Then your group should meet and attempt to sort each of the samples provided.

4. You have become familiar with the animals you have classified. You have observed them carefully to answer all the questions on the key. Your teacher is now going to provide your group with another organism to classify. Carefully follow each of the questions on your key to determine where the new macroinvertebrate falls in the key. When you have completed classifying your new organism, include its characteristics and name in your table.

5. When you followed the key, you had to make decisions about the characteristics you saw. Be prepared to share your decisions with the other groups in your class.

Why Classify Living Organisms?

Scientists have determined that all living things on Earth, past and present, share the following characteristics:

- They are made of units called cells.

- They get energy from the environment.

- They grow and develop.

- They reproduce.

- They respond to changes in the environment.

Living things are also different from each other. The cell is the basic structure and function of living things. Some living things are made up of only one cell. That single cell can do everything it needs to stay alive. Some living things are made up of many cells. You are made up of about 100 trillion cells!

Living things are different in other ways as well. Some living things, such as plants, can make their own food. Plants can also be divided into many smaller groups. For example, some plants have flowers.

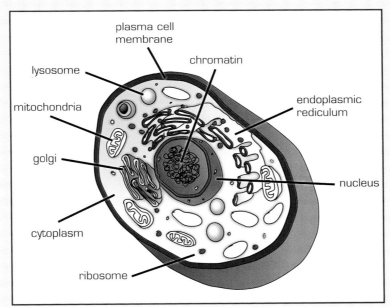

Classifying living things takes a lot of time and hard work. Scientists called **taxonomists** create systems for classification. They separate the millions of **species** of living things on Earth based on their differences and put them into groups. The systems they create help scientists communicate with one another.

History Connection

Carolus Linnaeus (1707–1778) is credited with developing a classification system based on an organism's characteristics. Linnaeus classified all organisms into two kingdoms, Animalia and Plantae. There have been many variations of this classification scheme throughout history.

species: a group of organisms that look alike and can breed with other members of the group and produce fertile offspring.

taxonomist: a scientist who classifies organisms by characteristics.

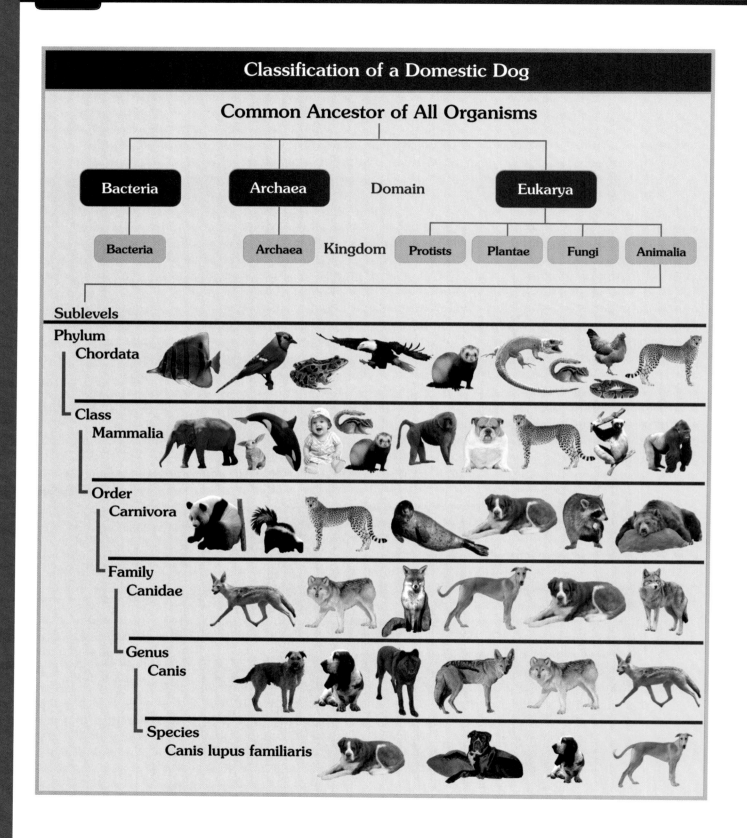

Classification of a Domestic Dog

Common Ancestor of All Organisms

Domain: Bacteria | Archaea | Eukarya

Kingdom: Bacteria | Archaea | Protists | Plantae | Fungi | Animalia

Sublevels

Phylum
 Chordata

Class
 Mammalia

Order
 Carnivora

Family
 Canidae

Genus
 Canis

Species
 Canis lupus familiaris

The method you used to identify your macroinvertebrates is called a **dichotomous key**. The key has a series of choices. Each choice leads to a different path through the key. If each choice is made carefully, the organism can be accurately identified and named.

Look at the chart on the previous page. Notice that the largest grouping is the Domain. The next largest is the Kingdom. The groupings get smaller and smaller until the organisms within a group are very similar to one another. The smallest grouping is the Species.

The number of kingdoms into which organisms are classified has varied over time. Scientists are constantly finding new living organisms that they didn't know existed. Because of this, the classification methods they are using to create order in the groups of animals are changing.

dichotomous key: a key used to identify living things.

What's the Point?

Scientists attempt to bring order to the many different organisms that live on Earth. To do this, they sort, or classify, living things based on their characteristics. Often, physical characteristics are used to classify animals. In the activity in this section, you observed animals and sorted them based on their physical characteristics. Scientists often use tools like a dichotomous key to assist them in classifying different organisms. Taxonomists continue to update these classification systems because new species are being found all the time. Even humans fit into the classification system of living things.

What characteristics might you use to classify this dragonfly?

3.3 Explore

The Marry Martans River Mystery: Macroinvertebrates in an Ecosystem

ecologist:
a scientist who studies the relationships between organisms and their environment.

You watched a video of scientists collecting macroinvertebrates. You should now have a good sense of how scientists organize and classify macroinvertebrates. Once scientists identify macroinvertebrates in an ecosystem, they can use this information to better understand the conditions in an ecosystem.

You also learned about diversity and abundance. Recall that diversity refers to the types of organisms found in an environment. Abundance refers to the number of each type. In this activity, you will examine the diversity of macroinvertebrates in an area. You will see how diversity can indicate water quality and ecosystem health. You will be working with some macroinvertebrate data collected by an **ecologist**. The ecologist has been asked to help the residents of a small community solve a mystery. What you learn from this case study will help you address this *Learning Set*'s question.

Examine a Case Study

A group of residents live on a small lake called Marry Martans Lake. The Marry Martans River flows into the lake at one end. The lake drains back into the river at the other end. (See the picture on the next page.) Over the past few months, the residents have noticed a lot of algae growing in the lake. The young people in the community know about water-quality indicators from their science classes. They remember that sudden algae and plant growth could be a sign of high amounts of fertilizer running off into the river.

The young people and their parents decide to investigate the case. Where might the fertilizer be coming from? They discover that there are three farms upriver. These farms are upstream from the lake and border the river. They wonder if fertilizer runoff from the farms is causing the problem. The residents discuss this with the farmers. Each of the three farmers denies that they have a fertilizer-runoff problem.

Each farmer knows that the river and the land are important to the community. They also know that the land use is needed for their business. Each farm claims to have safeguards in place that are designed to make sure harmful runoff does not enter the river.

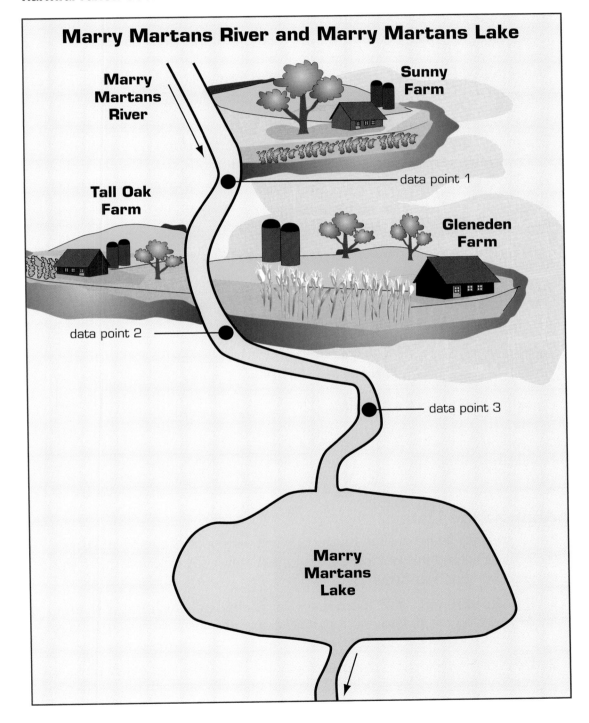

Marry Martans River and Marry Martans Lake

Marry Martans River

Sunny Farm

data point 1

Tall Oak Farm

Gleneden Farm

data point 2

data point 3

Marry Martans Lake

The residents contact a local ecologist. They explain their problem. The ecologist knows that there are twelve macroinvertebrates commonly found in this river. Some of the macroinvertebrate populations can be harmed by pollution.

The ecologist's team decides to collect macroinvertebrates from the three locations in the river identified on the map on the previous page. The locations are upstream from the lake. Each data collection point is located at the end of the runoff zone for one of the farms. These points are the last places where runoff from each farm would enter the river.

The team uses nets to collect macroinvertebrates from each location. They use a method like the one you watched in the video. Then each macroinvertebrate is identified using the dichotomous key. The table shows the different macroinvertebrates the ecology team collected at each point.

Location	Macroinvertebrates Collected		
data point 1 (Sunny Farm)	• midge larva • blackfly larva • leech • lunged snail	• cranefly larvae • sow bug • crayfish • dragonfly larvae	• mayfly larvae • caddisfly larvae • gilled snail • riffle beetle
data point 2 (Tall Oak Farm)	• midge larva • blackfly larva	• leech • lunged snail	
data point 3 (Gleneden Farm)	• midge larva • blackfly larva • leech • lunged snail	• cranefly larvae • sow bug • crayfish • dragonfly larvae	

Analyze the Data

See if you can figure out the mystery. Work with your group to analyze the data table and answer these questions. Later, your class will meet and discuss everyone's answers.

1. If polluted runoff could harm macroinvertebrates, which farms seem to be harming the macroinvertebrates? Support your answer with data and science knowledge.

2. Does the data support the claim by any of the farms that their pollution-control measures are working? Explain your answer with data and science knowledge.

The substances that farmers use on their fields can end up in local bodies of water.

3. List the macroinvertebrate species that were common to all three data points. If pollution is in the river, what effect do you think the pollution has on this set of macroinvertebrates?

4. Which species of macroinvertebrates were found at only one data point? Which point was this? If pollution is in the river, what effect do you think the pollution has on this set of macroinvertebrates?

The ecologist reviewed the data and believes that there may be more pollution entering the river than the farmers know. The ecologist visits each of the three farms with an expert from the Environmental Protection Agency (EPA). The EPA is a government agency that oversees and regulates how our society interacts with the environment. The EPA expert knows a lot about farm pollution.

The ecologist and EPA expert inspect the pollution-control measures that each farm is taking. Their findings are shown in the table on the next page.

Location	Pollution Control Analysis
Sunny Farm	Farm is adding very low levels of pollution to the river from runoff. In fact, the farm is below the limit allowed by the EPA.
Tall Oak Farm	Pollution-control measures are not working properly. Farm owner was unaware of gaps in the runoff system. Farm is contributing very high levels of pollution to the river from runoff. Runoff levels are above the limit allowed by the EPA by 200% (3 times what is allowed).
Gleneden Farm	Pollution-control measures are working for the most part. One gap in the system is a major source of runoff pollution. Runoff levels are above the limit allowed by the EPA by 15%.

5. Earlier, you decided which farms might be polluting the river. How does the data from the EPA report match up with the data collected at points 1, 2, and 3?

6. What does the earlier macroinvertebrate data suggest about the ability of macroinvertebrates at data point 2 to live in polluted water?

7. Look at data point 1. There are several macroinvertebrates that only appear at data point 1, but not at data points 2 and 3. What does the data from the EPA suggest about the ability of those macroinvertebrates to live in polluted water?

Communicate Your Ideas

Once all of the groups have completed these questions, your teacher will lead you in a discussion. You will review the answers and ideas generated by your group. Your teacher will ask you to group the macroinvertebrates from this activity into one of three groups.

Tolerant	Macroinvertebrates that can survive in polluted water.
Mildly Tolerant	Macroinvertebrates that can sometimes survive in polluted water.
Intolerant	Macroinvertebrates that cannot survive in polluted water.

Biotic Indicators

In *3.2*, you learned that one way to classify organisms is by their physical features. Those features evolved over very long periods of time. Animals have those specialized features because those features allowed previous generations of that animal to survive. These previous generations reproduced and had offspring with those features. Thus, the features keep being passed on to the next generation.

For example, certain macroinvertebrates have wings that allow them to avoid being easily captured by other organisms that try to eat them. These macroinvertebrates pass this trait onto their offspring. The feature remains important in the organism. These features might help animals avoid capture. They may also help them hunt food, better digest nutrients, camouflage the organism, or help them build a habitat (nest or den, for example). These features, or survival tools, help protect populations of organisms.

You know, however, that these protections or features are not a guarantee that an organism will avoid death. You probably are aware that, in nature, some organisms are the food source for other organisms. If you begin to see a decrease in the number of organisms in a population, then it could be a sign that the organism is being hunted and captured by other organisms in the ecosystem. You will learn more about this in a later section.

One other interpretation of a decrease in organism population is that the environment changed. In *Learning Set 2*, you read about how changes in water quality can be very harmful to the health of some organisms. Macroinvertebrates are one set of organisms that serve as a **biotic** (or living) **indicator** of water quality and health of the ecosystem.

biotic indicator: an indicator that an organism is or was alive (could be a fossil).

These animals serve as biotic indicators of the health of their aquatic ecosystem.

What's the Point?

Some of the residents near Marry Martans Lake noticed a sudden increase in the growth of algae. They worried that the farms had runoff problems. Macroinvertebrates are an example of living things that can serve as a biotic indicator of water quality. Explosion in growth of algae and plant life can also indicate a change in water quality due to fertilizer runoff. The residents were able to use the macroinvertebrate data to support their ideas about pollution in the Marry Martans River.

In this activity, you could see that when pollution entered the river, the variety of macroinvertebrates in that area decreased. Macroinvertebrates that are tolerant of pollution can survive new pollutants being added to the water. Intolerant and mildly tolerant macroinvertebrates will not survive when water becomes polluted. This is another way that scientists often classify organisms. Organisms can be grouped by how they react to changes in their environment.

Algal growth can make water unlivable for some macro- and microinvertebrates.

3.4 Investigate

Effects of Turbidity on Living Things

You learned how macroinvertebrates serve as an indicator of water quality in an ecosystem. That is just one example of how living things can serve as an indicator of water quality. Recall that in *Learning Set 2*, your class investigated how high levels of nitrates and phosphates can affect plant growth. Nitrates and phosphates increase the growth of plants. Plants can serve as an indicator of nitrates and phosphates in the water.

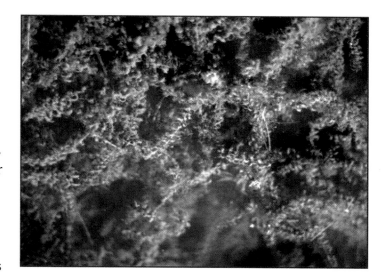

You also read about turbidity in *Learning Set 2*. Turbidity is a measure of how easily light passes through the water in a river or stream. Land development is common in many ecosystems. The soil that runs off the land often ends up in the river. The particles of soil mix in the river. This makes the water turbid. If turbidity is high, the water appears very murky and cloudy. Light cannot easily pass through highly turbid water. But what does this mean for living things in the water and the ecosystem? How might water plants be affected by highly turbid water?

You will work with your group to investigate this issue. You are going to study the effect of sunlight on the life of an aquatic (water) plant called elodea (ih-LODE-ee-uh).

Elodea is found naturally in many streams and rivers. It is found in North America and other continents. Elodea is easy to grow. Because of this, it is a plant commonly kept in aquariums. It is also often used in the laboratory to investigate how plants function. The question you will investigate is *How do plants react to changes in the amount of sunlight they receive?*

Run the Investigation

You will work with your group to investigate how the amount of light received affects elodea, but the results of your investigation can be applied to land plants as well.

Elodea is an aquatic plant that lives in freshwater streams. You will use four sprigs of elodea in your investigation. You will expose two sprigs to sunlight, and keep the other sprigs out of the sunlight.

Keeping the sprigs out of the sunlight would be similar to the conditions in a high-turbidity stream or river where very little sunlight is reaching the Elodea. You will record observations of what happens to the plants at the start of the investigation and again after thrity minutes.

The results of your investigation should help you answer the question *How do plants react to changes in the amount of sunlight they receive?*

Procedure: Set Up Your Investigation

Materials
- 4 sprigs of elodea
- 2 transparent plastic zipper bags (small)
- 2 shoeboxes, one with a lid
- water
- access to a lamp or a light tray
- *Elodea Investigation* page

1. Your teacher will provide you with the materials in the list. You will have about ten minutes to set up your experiment.

2. Label two plastic zipper bags with your name.

3. Take four sprigs of elodea. Use scissors to trim the stem end of each sprig. Pull the leaves off the bottom of each stem so there are no leaves in the first 1 cm (1/3 inch). Crush the stem end of each sprig with your fingers.

4. Put two sprigs of elodea in each sealable bag. Your teacher has added a half-teaspoon of baking soda to each bag. Add enough water to each bag to cover the plant. Seal the bags. Make sure the water is not leaking.

5. Carefully observe the plants inside the bags. Record your initial observations on an *Elodea Investigation* page similar to the one shown on the next page.

6. Put one bag with elodea *inside* a shoebox. Put the shoebox under the lamp or in the light tray, if available.

7. Put the other bag with elodea *inside* another shoebox. Place the lid on the shoebox. Put the shoebox under the lamp or in the light tray, if available.

8. Leave the shoe boxes undisturbed for thirty minutes.

9. What do you think will happen during your experiment? Record your prediction on your *Elodea Investigation* page.

Elodea Investigation

Name: _____ Date_____

Prediction

Describe what you think will be the effect of keeping elodea out of the light. Also, what difference do you think you will see between the two plants at the end of the investigation?

Observations

Record details of how the plants look and what you see in the bags. Draw a sketch of the elodea to help you describe what you see.

With Light	Without Light
Before	Before
After	After

What did you notice in the bag of elodea that was exposed to light?

_____

What did you observe in the elodea that was kept in the dark?

_____

What do you think caused the changes you observed?

_____

_____

● **You will return to this investigation later during class.**

Understanding Photosynthesis

You may remember from previous science classes that plants require certain things in order to grow. Recall the duckweed in *Learning Set 2*. You placed the plants in water. Then you added fertilizer (nutrients) to the water and placed the plant under the growth lights. Each of these factors was important for the growth of the plant.

Photosynthesis refers to a complex process that keeps plants alive and helps them grow. It takes place in plants, algae, and some bacteria. These organisms use water, **carbon dioxide**, nutrients, and light to survive and grow. They get these things from their environment.

photosynthesis: the process in which sugars are produced by plants and other organisms using water, carbon dioxide, and energy from sunlight.

carbon dioxide: a colorless, odorless gas commonly found in air; carbon dioxide is used by plants in the process of photosynthesis.

LIVING TOGETHER

glucose: a type of sugar. It is the main source of energy for living organisms.

sugar: a chemical compound produced by plants during photosynthesis. Sugars provide a source of energy used by living organisms.

oxygen: a colorless and odorless gas produced by plants during photosynthesis and used by animals for respiration.

cell: a basic unit of all living organisms.

chlorophyll: a green substance found in plants. Chlorophyll is used to capture energy from sunlight during photosynthesis.

chloroplast: the part (organelle) of a plant cell that specializes in photosynthesis. It contains chlorophyll.

They produce **glucose** (a **sugar**) and **oxygen**. Glucose is used as an energy source. The energy is needed to power the plant's life processes. Life processes include activities such as growing new tissue and reproduction. The oxygen produced is released into the environment.

Light is essential for photosynthesis. If plants cannot get enough light, they cannot produce the glucose they need to grow. The light can come from the sun or from indoor lights.

In plants, photosynthesis occurs mainly in the leaves. The leaves are made up of **cells** that contain a green substance. The substance is called **chlorophyll**. It captures the energy of sunlight. Chlorophyll is found in the **chloroplasts** of plant cells. To function properly, plants and algae also need small quantities of mineral nutrients, such as nitrate and phosphate. Land plants get these nutrients from the soil. Aquatic plants get them from the water.

All organisms need a source of energy to function. Animals eat plants and other animals to make energy. Plants make their energy through photosynthesis.

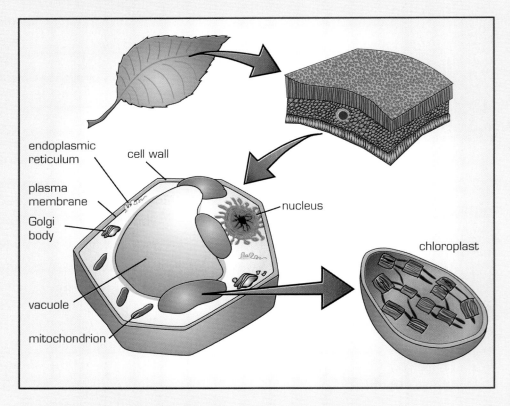

endoplasmic reticulum

cell wall

plasma membrane

Golgi body

nucleus

vacuole

mitochondrion

chloroplast

Project-Based Inquiry Science

cell respiration: a chemical process in which the energy stored in sugar is released. This process occurs inside the cells of plants and animals.

The processes of photosynthesis and respiration are connected. The oxygen produced by plants is used by animal cells during respiration. The carbon dioxide produced during respiration is released into the environment. This carbon dioxide is taken up by plants to photosynthesize. Animals and plants are therefore tightly linked. Each needs the other, and both rely on the environment to survive.

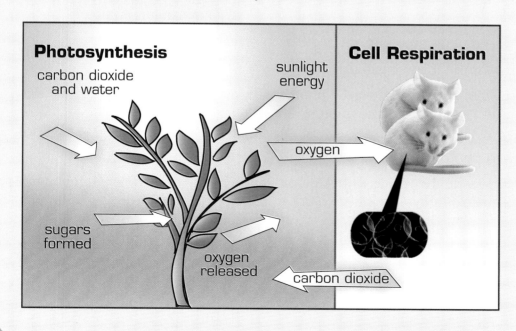

Photosynthesis

carbon dioxide and water

sunlight energy

Cell Respiration

oxygen

sugars formed

oxygen released

carbon dioxide

Observation and Recording Data

1. After about thirty minutes, return to your samples. Gently remove the lid and compare the contents of the two bags.

   Record any differences you observe in the bags.

   • What changes occurred in the bag of elodea that was exposed to light?

   • What changes occurred in the bag of elodea that was kept in the dark?

   • Are there any differences between the two bags? (Think about what happens during photosynthesis and see if you can spot any differences in what photosynthesis produces.)

2. Record your observations on an *Elodea Investigation* page. Although you are working on the investigation as a group, each member of the team should record his or her own observations. Later you will discuss your results among your group and compare them to those of other groups in the class.

Analyze Your Data

Once you have collected your data, discuss the results of your experiment with other members of your group. The questions below should help you organize your discussion. Take notes of what others are saying during the discussion.

   • What did you notice in the bag of elodea exposed to light?

   • What did you observe in the elodea that was kept in the dark?

   • What do you think caused the changes you observed?

Answer the following questions.

1. What is produced during photosynthesis? Write down at least two things that are the result of photosynthesis.

2. How do you think the investigation you conducted with elodea is related to photosynthesis?

3. How do the results of this investigation help you answer the investigation question *How do plants react to changes in the amount of sunlight they receive?*

4. What do the results of your investigation suggest will happen to water plants in a river that has very high turbidity? How might that affect other living things in the ecosystem?

Explain

Decide as a group on the best explanation for what happened. Remember, the best explanation is one that is best supported by the evidence. Write down your group's explanation. Be prepared to share your conclusions and explanation with your class.

In an *Investigation Expo*, share with the class your analysis of your data, your answer to Question 4, and your explanation. Listen carefully as other groups share their findings and explanations.

Demonstration

Your teacher may present to your class the results of another investigation with elodea. This investigation was done while you were running yours.

You observed what happened to the elodea over 30 minutes. Your teacher also used an instrument to measure the changes observed in the two elodea samples. You only collected visual observation data.

Discuss the results of your teacher's investigation with your class. Use the following questions as a guide:

- How do the results of your teacher's investigation compare with yours?
- Do the results from your teacher's investigation help you to better understand what was happening to the elodea that was exposed to light?

Revise Your Explanation

As a class, revise your explanation of what caused the changes you observed. Base your revised explanation on the results of your teacher's investigation.

Reflect

Answer the following questions. Be prepared to discuss your answers with your class.

1. How might changing the other influences on plant growth, like amount of water or nutrients, affect plants? Compare these changes to changing the amount of sunlight a plant receives.

2. High turbidity can cause changes in a plant's ability to photosynthesize. What other water-quality indicators might signal a problem for photosynthesis in plants?

3. What could happen to the other living things in the ecosystem if a plant could not complete photosynthesis? Provide an example or draw a diagram to describe what might happen.

Update the *Project Board*

As a class, update the *Project Board*. Include what you have learned about the effects of light on plants from your investigation. Include your explanations about what was happening in the *What are we learning?* column of the *Project Board*. Make sure you support your conclusions with evidence you have collected. Fill in the columns *What are we learning?* and *What is our evidence?*.

What's the Point?

Both animals and plants need a source of energy to survive. In this section you have investigated the effect of light on plants. You discovered that plants exposed to light were able to carry out photosynthesis and produce oxygen and sugars. Without light, plants cannot produce the sugars needed to grow.

In the next *Learning Set,* you will further investigate the relationships between animals, plants, and their environment. You will continue to build your understanding of how water quality can affect the ecosystem.

3.5 Explore

Connections to Other Living Things

You looked at how small organisms and plants in an aquatic ecosystem can be affected by changes in water quality. It might seem obvious that organisms that interact with the water would be affected. The question for this *Learning Set*, however, is how water quality affects living things in an ecosystem. So, the important question to consider now is *How might the effects of water quality on a few living things affect all of the living things in the ecosystem?*

To get a sense of how connected other living things are to one another, you can look at your own interaction with living things. Earlier, you learned about the needs of living things. You know that most living things require food or nutrients in order to survive. Consider exactly where you get your nutrients.

Procedure

1. Think about a simple breakfast of cereal and milk. You can buy cereal and milk at a grocery store. But where does it actually come from? With your class, create a diagram like this one shown on the right. Trace the parts of the breakfast back to their sources.

2. With your group, work together on another example. Examine a meal that some students your age enjoy, a cheeseburger and fries.

Break apart the components of this meal. Trace each part back to its source. Use the photo to help you identify all the different parts. Record your group's analysis on a sheet of paper. You might want to use a pencil. That way, you can erase mistakes or make changes as you break down the meal.

Communicate Your Ideas

Idea Briefing

Once your group has reached agreement, prepare a poster to share your analysis with your class. Your teacher will collect all the groups' posters and display them together. As a class, you will compare and contrast the various ideas and discuss the analyses.

Be sure to look for differences between analyses. Think about the following questions as you discuss the posters:

- Are there items that a group or groups forgot to include?
- How did groups break down the food differently?
- What trends do you see in the way students identified the sources?
- Which organisms seem to eat only plants? Which organisms seem to eat meat?

Food Chains

Organisms in a Food Chain

Recall that ecology is the study of how communities of plants, animals, and humans interact with each other and the physical environment. The activity you just completed points out one aspect of ecology. Organisms can be connected in a **food chain**. A food chain is a path of connected organisms where one organism relies on another as a food source. Look at a food chain that uses the cereal and milk example.

In this food chain, the grass grows because of photosynthesis. The cow eats this grass. The nutrients from the grass allow the cow to produce milk. Humans drink the milk.

You may have noticed that in both the cheeseburger example and the cereal and milk example, everything in the diagram always leads back to plants. Plants use light from the Sun during photosynthesis to make their food.

You may have also noticed that some non-plant organisms rely on other non-plant organisms as sources of food. Each of the organisms in the food chain can be classified by

• its location in the food chain, and

• the role it plays in the food chain.

Some organisms are **producers**. Producers are organisms that are capable of making their own food. Plants and some bacteria make their own food. They are producers. The grass in the food chain to the right is a producer.

food chain: a sequence that shows what eats what in an ecosystem.

producer: an organism capable of making its own food.

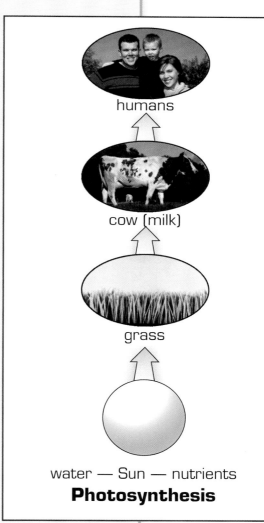

humans

cow (milk)

grass

water — Sun — nutrients
Photosynthesis

Some organisms are **consumers**. Consumers cannot make their own food. They take in food by consuming other organisms. Food chains can contain several different types of consumers.

- primary consumers eat producers,
- secondary consumers eat primary consumers, and
- tertiary (TER-shee-air-ee) consumers eat secondary consumers.

In the food chain shown on the previous page, the cow is a primary consumer and the human is a secondary consumer.

Consumers can also be classified as herbivores, carnivores, and omnivores. A **herbivore** is a consumer that eats only plants. Examples include cattle, horses, some insects, koala bears, and elephants. A **carnivore** is a consumer that eats mainly the meat of other organisms. Examples include hawks, eagles, snakes, spiders, and sharks. An **omnivore** is a consumer that eats both plants and the meat of other organisms. Examples include pigs, bears, some rodents, humans, and foxes.

consumer: an organism that must get its food by eating other organisms.

herbivore: an organism that obtains its food only from plants.

carnivore: an organism that obtains its food only from other animals.

omnivore: an organism that obtains its food from plants and animals.

predator: an organism that hunts and kills other organisms.

prey: organisms that are hunted and killed by other organisms.

Also, carnivorous consumers (organisms that are carnivores) are known as **predators**. These organisms hunt and kill other organisms, which are called **prey**. Prey does not also have to be a carnivore. Prey can be a herbivore like an insect. However, the organism that hunts and eats it is always a predator. The photo at right shows a predator, the hawk, and its prey, the mouse.

Food Chains and Energy

Your cheeseburger diagrams are not complete food chains. You might have identified items that are only parts of an organisms as consumers or producers.

For example, you might have identified "tomato" and then drawn a line to "tomato plant." Food chains connect organisms that are producers and consumers. Tomatoes do not consume tomato plants. However, your diagram helped you identify the parts of the meal and analyze their sources. Some complete food chains for the cheeseburger meal might look like the ones on the right. Notice how the arrows point *up* the food chain, toward the consumer.

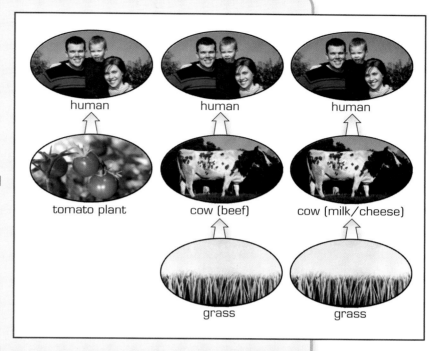

The arrows in a food chain show connections. When you examine the whole set of connections in each food chain, you can learn something very important. Remember earlier you saw that most food chains start with a plant that grows by photosynthesis. The plant takes energy (in the form of light) from the Sun. This energy is used to produce glucose (a type of sugar). The glucose is stored in the material that makes up the plant. It is stored in its stalk, its leaves, and its flowers.

In the cheeseburger food chains, the glucose is stored in the blades and seeds of the grass. The cow eats the grass. Next, it digests the grass. (Digest means to break down food so the body can use it.) The glucose is then passed on to the cow. The glucose and other nutrients are used to help the cow grow. They are stored in the tissues that make up the cow. When humans eat the hamburger, they digest the meat of the cow. By doing so, they obtain the nutrients in the meat.

The matter that makes up each member of the food chain is passed from the producer on to the highest-level consumer. This is a very important idea in understanding how living things are connected and rely on each other. The energy from the Sun that was processed by the plant is passed on to every member of the food chain as each member of the food chain is consumed. People would not have anything to eat without the Sun's energy.

Stop and Think

Use what you have learned to answer the following questions.
Be prepared to discuss your answers with the class.

1. Identify all of the producers and consumers in your diagrams.
   Which consumers are primary, secondary, and tertiary?

2. Identify the organisms that are herbivores, carnivores, and omnivores.

3. List all of the predators in your diagram. List all of the prey.

4. Explain how it is possible to be both a predator and prey.

What's the Point?

Using the diagrams you created, you were able to identify producers, consumers, carnivores, omnivores, and predators and prey. These food diagrams show you how these organisms are connected.

Food chains are very important in understanding the health and condition of an ecosystem. The living things in an ecosystem rely on other living things as food sources. Energy from the Sun is, essentially, transferred up the food chain. This is a very critical aspect of understanding the ecology of a community. In the next section, your class will investigate changes in food chains and how these changes could affect the living things in an ecosystem.

Bears are omnivores.

3.6 Investigate

Modeling Changes in Food Chains

You have investigated how organisms are connected in food chains. When you drink milk, you become part of a food chain that connects you, a cow, the grass, nutrients that help the grass grow, and the Sun.

You learned that consumers need to get their food from other organisms. Therefore, all organisms are part of a food chain. On one side of a link in a food chain, an organism is connected to the organism it consumes. On the other side of the link, the organism is connected to the organism that consumes it.

These things are connected in a food chain.

In this section, you will be looking more closely at how the interactions in food chains work. Instead of looking at individual organisms, you will investigate how many individuals from different groups within a food chain interact. In the real world, there are many individuals of each type of organism. There are many blades of grass that a cow will eat. There are also many cows that produce milk. There are many humans who drink the milk produced by the cows.

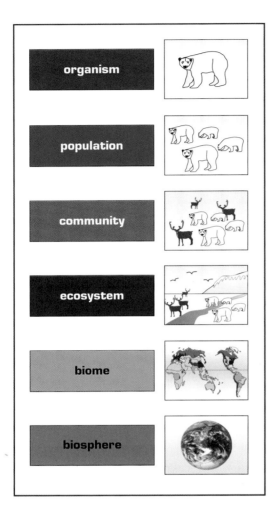

A group of individuals of the same species that live in a specific location is called a **population**. Populations of different organisms living in an area form a **community**. To understand how the organisms depend on one another and the environment in a community, scientists look at how their populations change over time. Because the interactions are difficult to observe and keep track of in the real world, scientists use computer models to simulate them. You will use a computer program called NetLogo to simulate population changes in a model community, like scientists do.

Design a Population Model

Imagine a model community that consists of a large area of grassy land where a population of mice lives. On the same land, there is also a population of coyotes. The grass, mice, and coyotes form a community.

With your group, discuss how the grass, the mice, and the coyotes are related in a food chain. Which are the primary producers in this food chain? Which are consumers? What eats what in this food chain? As a group, you need to come to an agreement about how this food chain works.

population: a group of organisms of the same species that live in a particular location.

community: the populations of different organisms living in a particular area. Organisms in a community depend on one another for survival.

Think about the populations of these organisms that might be a model community. How much grass do you think they need? How many mice and coyotes do you think can live together in the area?

Each individual organism goes through a life cycle during which it grows, reproduces, and dies. Think about the factors that might affect the growth, reproduction, or death of each organism that will be in your model.

How might the factors affecting an individual cause the population to change over time? With other members of your team, create a list of factors that might affect the populations of grass, mice, and coyotes.

The answers to the following questions will help you create the list:

- What might cause the amount of grass to change? List all the factors that might increase or decrease the amount of grass in your model community.

- What might cause the numbers of mice or coyotes to change? List all the factors that might increase or decrease the number of mice or coyotes in your model community.

Draw the food chain you imagine. Then record the factors that affect each population in your *Model Population Predictions and Observations* chart.

Model Population Predictions and Observations

Name: _____ Date: _____

1. Draw a diagram of the food chain that links the grass, mice, and coyotes in the space below.

2. List the factors that you think affect each population.

	Factors that might increase the population	Factors that might decrease the population
grass patches		
mice		
coyotes		

3. Record the data you collect when you run the model with Baseline Conditions. The first row of data has been entered for you.

Baseline Conditions			
Generations	Number of grass patches	Number of mice	Number of coyotes
0 (start)	1412	205	70

4. Record the name of your scenario. Write down your prediction about what will happen to the number of grass patches, mice, and coyote. Then record the data you collect when you run the model.

Baseline Conditions			
Generations	Number of grass patches	Number of mice	Number of coyotes
Prediction			
0 (start)			

How to Use NetLogo

You will be using a computer software program to investigate how organisms connected in a food chain can be affected by changes in the population of one organism. Open the program in your computer. Your teacher will tell you which model to open.

Once the model is loaded, you should see a screen like the one on the next page.

The screen is divided into three main sections:

- The *model input window* is on the left.
- The *graphics window* is on the right.
- The *plot graph window* is on the bottom.

The *model input window* is where you control the populations of grass, mice, and coyotes. You can slide the red bars on the blue slides to change the **parameters** for each organism.

parameters: the limits or boundaries (in this case, values).

The population of grass is controlled by the *grass settings* slide. Moving the slide left or right lets you choose how much of the land is covered by grass.

The populations of mice and coyotes are controlled by three different slides:

- *number-of-mice* and *number-of-coyotes*. Move the slide bars left or right to select how many individual mice and coyotes will be in your community at the start of your simulation.

- *max-mouse-age* and *max-coyote-age*. This is how long each individual mouse or coyote can live. Slide the bars of these slides to choose a maximum lifetime for each type of organism.

- *max-mouse-offspring* and *max-coyote-offspring*. Slide the bars of these slides to select the number of offspring for each type of organism.

You will need to adjust the parameters for grass, mice, and coyotes to start. Once you have selected the values of each parameter, click the **<setup>** slides (near the top). On the right of the screen you will see a graphic representing your model community.

Move the red bar on the *speed adjustment* slide above the graphic to the center, as shown in the picture.

Press the **<go/stop>** slide to start the simulation. As the model runs, the graphic on the right will change to show how the populations of grass, mice, and coyotes change over time. The changes in population over time for each organism are also displayed in the plot graph window on the bottom.

You can stop the simulation by pressing the **<go/stop>** slide. Press the **<go/stop>** slide again to continue the simulation.

To run another model, change the initial parameters, and then press the **<setup>** and **<go/stop>** slide.

Run a Population Model

Now get ready to use the computer program NetLogo to simulate how the populations of grass, mice, and coyotes change.

1. Before you can use the program, you need some information about how it works. Read the section *How To Use NetLogo* to learn about the program.

2. Open the NetLogo model as instructed by your teacher. For the first simulation, all groups in your class will assume the same initial conditions for the model community. Your initial conditions are shown on this page.

3. Use the slide bars to adjust the parameters to the values on this page. These values represent a typical situation in a community of grass, mice, and coyotes.

4. Press the <**setup**> slide. Slide the *speed adjustment* bar toward the center. Observe the graphic that appears in the *graphics window*. Identify the objects in the graphic.

   a) What does each object represent?

5. Press the <**go/stop**> slide to start the simulation. Notice how the graphic on the right changes to represent changes in the populations after each generation. Let the simulation run for a while. Press the <**go/ stop**> slide after about 100 generations. Look at the generation counter on the left of the screen to know when to stop. Notice any changes in the graphic at the right.

   a) Does the amount of grass look the same as when you started? How has it changed?

   b) How has the number of mice and coyotes changed?

6. To find the answers to the questions you made predictions about, you will run the simulation again. You will use the same initial conditions you just used. But this time, during the run, you will collect data to analyze later.

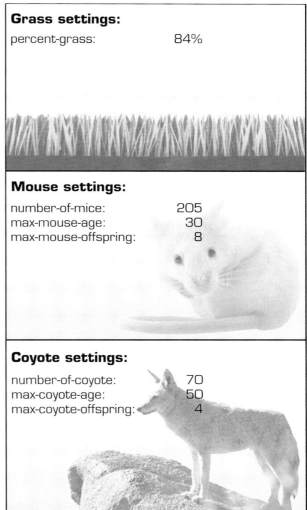

Grass settings:

percent-grass:	84%

Mouse settings:

number-of-mice:	205
max-mouse-age:	30
max-mouse-offspring:	8

Coyote settings:

number-of-coyote:	70
max-coyote-age:	50
max-coyote-offspring:	4

Model Population Predictions and Observations

Name: _____   Date: _____

1. Draw a diagram of the food chain that links the grass, mice, and coyotes in the space below.

2. List the factors that you think affect each population.

	Factors that might increase the population	Factors that might decrease the population
grass patches		
mice		
coyotes		

3. Record the data you collect when you run the model with Baseline Conditions. The first row of data has been entered for you.

Baseline Conditions			
Generations	Number of grass patches	Number of mice	Number of coyotes
0 (start)	1412	205	70

4. Record the name of your scenario. Write down your prediction about what will happen to the number of grass patches, mice, and coyote. Then record the data you collect when you run the model.

Baseline Conditions			
Generations	Number of grass patches	Number of mice	Number of coyotes
Prediction			
0 (start)			

Record the initial conditions for the parameters in your data chart.

7. Restart the simulation by pressing first the <**setup**>, and then the <**go/stop**> slides. Stop the simulation after 20, 40, 60, 80, and 100 generations. After each stop, record in your data chart the following numbers, as they appear in the screen:

- which *generation*—e.g., 0 (at start); 20; 40; 60; 80; 100

- *number of grass patches*—locate this number on the screen. Record the value shown when you stop.

- *number of mice*—locate this number on the screen. Record the value shown when you stop.

- *number of coyotes*—locate this number on the screen. Record the value shown when you stop.

Don't worry if you cannot stop the program exactly after 20, 40, etc., generations. Try to stop as close to those numbers as possible. But remember to record in the data chart the *generation* that appears on your screen when you stop.

Analyze Your Data

Discuss the data you collected during the simulation with other members of your group.

Answer the following questions during the discussion. Record the answers. Be prepared to share your answers with the class.

1. What happened to the amount of grass during the simulation? Support your answer with evidence from the data.

2. How did the population of mice change during the simulation? Use examples from the data you collected when answering the question.

3. How did the population of coyotes change?

4. Compare the change in the population of coyotes to that of the grass and mice in your data. Did the coyotes change more, less, or about the same as the grass and mice? Use evidence from your data to answer the question.

5. What relationships did you find among the populations of the different organisms? If you have a computer program available, you may want to create a graph of your results to see the relationships. The graph will help you to communicate your ideas better when you discuss the results with your class.

6. When did you observe periods during which one population is increasing while another is decreasing? Why do you think this might happen? Why do you think this happened?

7. Which population changed the least during the simulation? Why do you think a population might change only very little?

8. What do you think might happen to the populations in your model community in the future 100 generations? Describe what you think might happen using information from the data, and what you know about how the grass, mice, and coyotes are connected.

Population Changes in an Ecosystem

The number of individuals in a population changes over time. At any point in time, the size of a population depends on several things:

• the size of previous generations,

• the amount of food available,

• the number of offspring that survive, and

• other environmental factors.

Each individual in a population goes through a life cycle. It grows, reproduces, and dies. To grow and reproduce, each individual needs food for energy. The more individuals in a population, the more food they need. They get their food from the environment. Therefore, the size of a population affects what happens in the ecosystem.

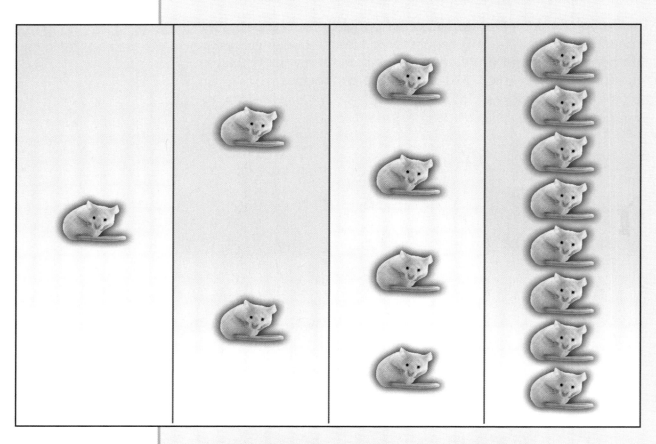

Project-Based Inquiry Science

Math Connection

If environmental conditions are right, and food is abundant, populations can grow very fast. For instance, a population of the bacterium Escherichia coli (E. coli) can double in size every 20 minutes. A population of 100 individuals of E. coli can grow to over 50,000 individuals in just three hours.

In your computer simulation, mice get their energy from eating grass. The more mice, the more grass they will eat. As the population of mice increases, the amount of grass decreases. The mice have a direct effect on the amount of grass. However, with less grass, fewer mice can get enough food to survive. Therefore, the population of mice starts to decrease after awhile. The population of mice has a direct effect on the grass. The grass has a direct effect on the mice.

Suppose every mouse in your model gave birth to two mice. The number of mice would grow very quickly. The rate of reproduction has a large effect on how a population changes over time. When there is lots of food, more individuals are able to reproduce. More young become adults. This will increase the population over time.

This did not happen in your simulation. In fact, it rarely happens in the real world. This is because not all individuals reproduce. Some die of natural causes before they reproduce. Some are consumed by predators before they have offspring. When food is scarce, some die of starvation before they reproduce.

Run a Population Model When the Ecosystem Changes

equilibrium: a condition of a system in which all influences cancel one another. The result is an unchanging, or balanced, system.

You have used NetLogo to model changes in the populations of grass, mice, and coyotes in a healthy ecosystem. When an ecosystem is healthy, population sizes change temporarily. They can increase or decrease over a few generations. But they tend to recover and become stable over time. Over many generations, the numbers of grass patches, mice, and coyotes reach **equilibrium**. These conditions represent a balanced ecosystem.

You are now ready to use NetLogo to simulate how populations in a community change when the conditions in the ecosystem change.

Procedure

1. Each group will model one of the six scenarios on the next pages. Read your assigned scenario. As you are reading, pay attention to the section describing the conditions that have developed. Look at how those changes correspond to parameter settings in your model.

2. Predict what will happen in your simulation based on what you learned about how this ecosystem works in your previous simulation. Record what will happen to the populations of grass, mice, and coyotes when conditions change as described in your scenario. Report your prediction in your data chart.

3. Select the initial parameters for the grass, mice, and coyotes.

4. Make sure the speed adjustment bar is set at mid-range. Press the <**setup**> and <**go/stop**> slides to start the simulation.

5. Stop the simulation after 20, 40, 60, 80, and 100 generations as you did earlier. Collect data for the number of grass patches, mice, and coyotes each time you stop. Record this data in your data chart, together with each generation number.

A woodmouse eating fallen sunflower seeds from a bird feeder in a garden.

Scenario 1
Drought

During the past two years, the patch of land in your model community has received very little rain. The lack of precipitation has caused drought conditions. There is much less grass than there used to be.

- Set the *percent-grass* in your NetLogo model to 40%.

- Set all other parameters to the values they had in the balanced ecosystem.

- Run NetLogo to see how drought affects the numbers of mice and coyotes.

Scenario 2
Another Prey for Coyotes

About a year ago, several farmers moved into the area near your model community. The farmers have set up chicken coops on their land. The coyotes in your community have discovered this new source of food. They visit the chicken coops often. With lots of food available, the number of coyotes in the population has gone up. Their rate of reproduction has also increased.

- Set the *number-of-coyotes* in your program to 100.

- Set the *max-coyote-offspring* to 6.

- Set all other parameters to their values in the balanced ecosystem.

- Run NetLogo to see what happens in a community when the number of predators goes up.

Scenario 3
Mice Find Another Source of Food

Some farmers living close to your model community recently started growing wheat on their land. The mice in your community prefer wheat to the grass. They have a new and abundant source of food. The rate of reproduction among the mice has increased. The number of young has also increased.

- Set the *number-of-mice* in your program to 400.

- Set the *max-mouse-offspring* to 10.

- Set all other parameters to their values in the balanced ecosystem.

- Run NetLogo to find out how well the ecosystem can support this large population.

Scenario 4
A Season of Abundant Rains

For about one year now, the patch of land where your community lives has received lots of rain. As a result, there is a lush carpet of grass covering the entire area.

- Set the *percent-grass* in your program to 100%.

- Set all other parameters to their values in the balanced ecosystem.

- Run NetLogo to see what is the effect of this abundant source of food for the mice and coyotes in the community.

Scenario 5
A Coyote-Virus Epidemic

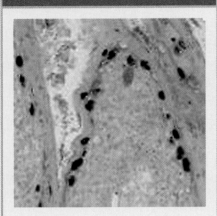

Over the past several months, a virus has infected the population of coyotes in your model community. First, some coyotes got the virus from drinking contaminated water. The infection then spread quickly among the population. Many adult coyotes have died. The number of offspring has also dropped dramatically.

- Set the *number-of-coyotes* in your program to 8.

- Set the *max-coyote-offspring* to 2.

- Set all other parameters to their values in the balanced ecosystem.

- Run NetLogo to see how a reduction in the number of coyotes affects the mouse population. Observe how the number grass patches changes too.

Scenario 6
Pesticides in the Environment

Large areas of land near your community are cleared to grow crops. To prevent bugs from destroying the crops, the fields are sprayed with **pesticides**. Unfortunately, these chemicals affect the young mice in your community. Many of them die before reaching maturity and reproducing.

- Set the *max-mouse-offspring* to 2.

- Set all other parameters to their values in the balanced ecosystem.

- Run NetLogo to see how a reduction in primary consumers affects the predator population and the number of grass patches.

Analyze Your Data

1. Compare the results of your scenario simulation with your prediction of what would happen. Was the prediction accurate? Why or why not?

2. Compare the results of your scenario simulation with the simulation of the balanced ecosystem you ran earlier. What changes in the populations have occurred?

3. What was the effect of the change in conditions in your simulation on the populations of organisms in your community? Do your best to explain why.

pesticide:
a chemical used to prevent bugs from destroying crops.

Communicate Your Results

Investigation Expo

Use the *Analyze Your Data* questions as a way to discuss the results of your simulation with your group.

Create a poster describing the conditions you were modeling in your simulation. Make the description as detailed as you possibly can. Include all the parameters you selected to run your simulation. Illustrate how the populations of mice, grass, and coyotes changed. Indicate on your illustration the relationships between the populations. Also include an explanation of how the changes in one population caused changes in the others.

During the *Investigation Expo,* you are going to explain how your model community worked. You need to include enough details in your presentation so that your class will understand how the changes in one population, or conditions in the environment, affected the other populations. Answer all of the following questions in your presentation:

1. How did the populations of grass, mice, and coyotes change in your simulation?

2. Was the amount of grass after 100 generations more than, less than, or about the same as at the beginning?

3. Did the number of mice and coyotes increase, decrease, or stay the same after 100 generations?

4. What changed in your new scenario compared to the balanced ecosystem?

5. Do you think the conditions of your simulation improved or worsened the health of the balanced ecosystem?

As you listen to the investigation presentations of the other groups, observe

LIVING TOGETHER

how the populations change for each scenario. Compare the relationships among the populations in the different scenarios. Are the relationships the same? Did any of the scenarios result in the disappearance of some organisms? How do you think that might affect the populations of the other organisms in the community?

What's the Point?

Populations of organisms vary over time because of interactions with other organisms in the food chain and because of changes in the environment. Given enough time to adjust, a community in an ecosystem reaches a balance where the populations of organisms fluctuate up and down a little bit but remain relatively stable.

This balance can be upset when conditions in the environment cause changes in one population. Because the organisms interact with one another and their environment, all other populations are affected as well.

Sometimes the changes introduced in the environment result in the disappearance of an organism. In the next section, you will continue to investigate the relationships among the organisms in a community. You will also examine the relationship between organisms and their environments.

Populations of animals in a healthy ecosystem change temporarily, but will reach equilibrium over time.

Project-Based Inquiry Science

3.7 Explore

Connections between Living Things in an Ecosystem

You investigated the effects of sunlight on the growth of plants. You learned how plants use energy from the Sun during photosynthesis to create food. The energy is used to make sugars. The plants use the sugars to function and grow. Plants are producers. They are the foundation of the food chain. Without the Sun, plants cannot grow. In your model food chains, you learned how populations can change when plant growth changes.

There are many organisms that live in a watershed along with plants. You will work with your group to explore how organisms in a watershed ecosystem rely on one another. Your teacher will provide you with a set of flash cards. Each card has a picture of a single aquatic organism. The card also contains information about the organism. You will use these cards to explore possible connections among these living things.

Within this quite country scene are many organisms that depend on a healthy aquatic eco-system.

Procedure: Simple Connections

1. With a partner, examine the flash cards. Read and discuss all the information on each card. The cards will help you think of possible connections between the organisms pictured. If you have any questions about the information, ask your teacher for more explanation.

2. Assemble a food chain using the cards. Place the cards on a flat surface (table or desk). Place connections between the cards using flat sticks with arrows. Once you have built your idea of the food chain, record your food chain on a *Food-Chain Records* page, like the one shown on the next page. Circle the organisms in the chain, and make sure the arrows point in the direction of the energy flow.

3. Meet with the other half of your group. Compare your cards and food

chains. Record their food chain on your *Food-Chain Records* pages in the space indicated.

4. As a group, combine your two food chains into one connected food chain. Be careful. There is a catch: *If both food chains have the same organism, your new configuration can only have that organism listed once.*

   For example, if you both have a card with a rabbit on it, you can only use one of the rabbit cards. As before, place the cards on a flat surface (table or desk). Place connections between the cards using the sticks. (Hint: It is possible to place more than one stick to and from an organism.)

5. Once you have built this larger food chain, record this third food chain on your *Food-Chain Records* pages in the space indicated. Circle the organisms in the chain, and make sure the arrows are pointing in the direction of the energy flow.

Food-Chain Records

Name: _____ Date: _____

Record your first food chain here.

List any of the following items that appear in your food chain:

Producers	
Primary Consumers	
Secondary Consumers	
Tertiary Consumers	
Predators	
Prey	

Record the food chain built by the other half of your group here.

List any of the following items that appear in your food chain:

Producers	
Primary Consumers	
Secondary Consumers	
Tertiary Consumers	
Predators	
Prey	

Record your group's combined food chain here.

List any of the following items that appear in your food web:

Producers	
Primary Consumers	
Secondary consumers	
Tertiary Consumers	
Predators	
Prey	

The Missing Link in the Food Chain: Decomposers

Consider what happens to the matter and energy after the final consumer. There is another type of role in the food chain in addition to producers and consumers. The third role is the **decomposer**. Decomposers break down the tissues of dead plants and animals. Bacteria and fungi are examples of decomposers that break down animal tissue. There are also animals known as **detritivores**. These include buzzards, flies, earthworms, and cockroaches. They feed on dead plants and animals.

In turn, other decomposers break down their bodies when they die. The role of the decomposer is to get rid of all the waste and tissues of dead plants and animals.

In the end, decomposers leave behind nutrients taken from the plant or animal tissue and leave them in the surrounding soil and water. These nutrients are then processed by plants during photosynthesis. The whole food chain begins again. Once again, energy from the Sun is passed on to other organisms.

The fungus growing on the dead tree trunk helps it decompose by breaking down the tissues of the dead tree.

decomposer: an organism that breaks down the wastes and remains of other organisms.

detritivores: organisms that feed on dead plants and animals.

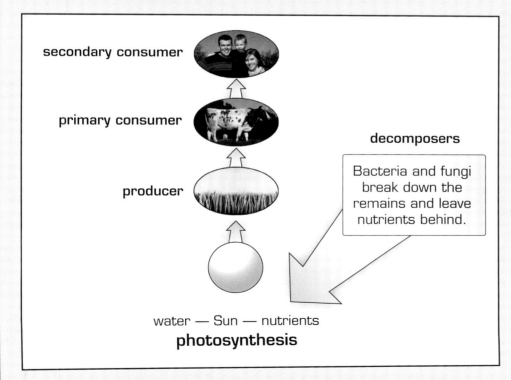

secondary consumer

primary consumer

producer

decomposers

Bacteria and fungi break down the remains and leave nutrients behind.

water — Sun — nutrients
photosynthesis

food web:
a series of
interlocking
food chains.
They show the
transfer of energy
through the
different levels in
an ecosystem.

Food Chains Connect to Form Food Webs

You may have noticed that your food chains look different. Up until now, your food chains have always connected one organism to another in a straight line. One organism follows the next in a single line. To connect the food chains in the activity you just did, it was impossible to keep everything in a straight line.

Food chains show how at least three organisms are connected. Suppose one of the organisms in that food chain consumes an organism not listed in the food chain. In that case, the food chain does not tell the entire story. The picture of how organisms interact with one another in an ecosystem is not complete.

Most animals are part of more than one food chain. They eat more than one kind of food to stay alive. Thus, food chains cross each other. These interconnected food chains form a **food web**. Food webs expand the food chain concept from a single line of organisms into a network of interactions.

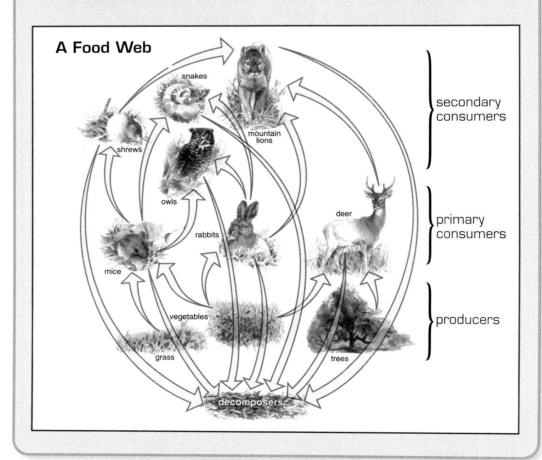

A Food Web

snakes

shrews

owls

mountain lions

mice

rabbits

deer

vegetables

grass

trees

decomposers

secondary consumers

primary consumers

producers

Procedure: Complex Connections

1. Your teacher will partner your group with another group.

2. Begin by recording the food chain of each on another set of *Food-Chain Records* pages. Using each group's cards and arrows, build one food web that combines the two small food webs on your table or desk. Once again, you may not use a card more than once from your set of cards — one organism, one card. To make longer connections (arrows), place the flat sticks end to end to reach across your surface.

3. Once your team has settled on a food web, record your team food web on your new *Food-Chain Records* pages in the space indicated. Then obtain two blank sheets of poster paper from your teacher. Each team will use these two sheets of paper to present their food web to the rest of the class.

4. On the first sheet, place your flash cards on the sheet (as you did on the table or desk). Securely tape them to the sheet. Instead of using sticks, use a marker to draw connections between organisms. Remember that the arrows should point toward the consumers in the direction of the energy flow.

5. On the second sheet of poster paper, identify

   - organisms that are producers;

   - which organisms are primary consumers, which are secondary consumers, and which are tertiary consumers;

   - the herbivores, the carnivores, and the omnivores in your chain; and

   - the organisms your group omitted from the food chain. Be sure to group these cards together and tape them to the second sheet of poster paper.

Be sure to write large enough so others will be able to read your posters during your presentation.

Communicate Your Ideas

Idea Briefing

Each group will have five minutes to share its food chains with the other groups in the class. Your teacher will lead you in a discussion of the various chains that groups created. You will discuss the producers and consumers you have identified and any possible organisms that your group omitted from the food chain.

While listening to the presentations, be sure to look for differences in food chains. Think about the following questions as you discuss the posters:

   - Are there items that a group or groups forgot to include?

- How did groups organize the organisms differently?
- What similarities and differences do you see among groups in the list of organisms that could not be incorporated into the food chain?

Reflect

Work with your small group to answer the following questions. Be prepared to share your answers with the class.

1. What patterns or trends did you notice across each of the food webs?

2. What similarities exist between all of the food webs?

3. What do you notice about the number of producers and tertiary consumers compared to the number of primary and secondary consumers? What does this comparison say about the energy required for tertiary consumers to survive?

4. List the possible connections you see between your team's food web and some of the other food webs presented.

5. Do you think there is a way to combine all these food webs into one single food web? Explain why or why not.

Update the *Project Board*

Recall that the question for this *Learning Set* was *How can changes in water quality affect the living things in an ecosystem?* Discuss with your class what you have learned about the biotic parts of an ecosystem. Record on the *Project Board* what you discovered about the relationships among these organisms.

What's the Point?

Food chains are important in understanding the condition of an ecosystem. There are often many different organisms that live in any ecosystem. Most are part of more than one food chain. Therefore, a food chain alone does not provide a complete picture. You cannot see all of the relationships among the organisms in a community just by looking at a single food chain.

To see all of the relationships among organisms, you need to look at all the feeding relationships among the organisms. In most ecosystems, they form a complex network. Ecologists call this network a food web. A food web connects all the food chains in an ecosystem. A food web provides you with a big picture that can help you understand how the effect of water quality on one organism can affect an entire ecosystem.

More to Learn

More to Learn about Earth's Biomes

In this Unit, you investigated the living and nonliving parts of a river ecosystem. There are many different types of ecosystems on Earth. To better understand the natural world, scientists have classified ecosystems into a small number of different types, called **biomes**.

Rivers, streams and most lakes are considered fresh water because they do not have salt in them. The freshwater ecosystem of rivers is one example of a biome. A biome is a community of plants and animals best **adapted to** an area's natural environment and climate. In the freshwater environment you have been studying, the plants and animals are well adapted to living there. You saw that in *Learning Set 3,* as you studied the different types of animals and plants that live in this biome.

Another word for a water environment is aquatic. **Aquatic biomes** are biomes that depend on water. Aquatic biomes can be freshwater like the river you studied, or they can depend on salt water. Freshwater and saltwater biomes are the most important biomes on Earth. Water is necessary to all life. All life on land depends on water for survival. Many species live in water for all or part of their lives.

biome: a community of plants and animals living together in a certain kind of climate.

aquatic biomes: biomes that depend on water.

adapted to: suited for living in a given environment.

World's Biomes

Besides aquatic biomes, there are also land, or **terrestrial**, biomes. Aquatic biomes are defined by the type of aquatic environment. Terrestrial biomes are characterized by the main type of vegetation covering a large area.

terrestrial: related to the land.

Look at the information below. Notice the different types of aquatic environments. You might have visited or seen pictures of these types of biomes. Think about the places you have visited or seen in pictures that might be these kinds of biomes. You have studied a river, which is a freshwater biome. What other freshwater biomes can you think of? Do you know of any saltwater, or marine, biomes?

Aquatic Biomes

Freshwater Biomes

Rivers and Streams

This type of biome is defined by bodies of fresh water that flow in one direction. They vary in size from small to large, and they are found all over the world.

Streams and rivers collect water from a watershed, and they normally grow in size from their source to their end. Water quality along a river changes depending on the particular land use of the areas the river flows through. The species of animals and plants that live in these biomes depend on the climate and the quality of the water.

Lakes and Ponds

These biomes are defined by bodies of water that fill a depression on Earth. They flow very slowly or not at all. They are found in all climates. Lakes and ponds vary in size from just a few square meters to thousands of square kilometers. Temperatures vary in ponds and lakes depending on the season, the location, and the depth.

The shallow waters of lakes and ponds receive a lot of sunlight and have an abundance of small plants and animals. The plants photosynthesize and support the entire community of animals in a lake's food web.

In very deep lakes, the bottom layers are dark and cold. With little sunlight available for photosynthesis, there are no algae, and only a few animals that feed on organic waste can survive there.

Wetlands

Wetlands are areas that are under water for at least part of the year. They are found in all climates. The vegetation in wetlands can survive flooding. Because the water and vegetation can offer shelter and protection, wetlands support a large diversity of plants and animals. Plants found in wetlands include both grasses and trees. In wetlands close to ocean coasts, the water often contains a lot of salt. Only animals and plants that tolerate high levels of salt can survive in this environment.

Marine Biomes

Shorelines

This type of habitat occurs where oceans and land meet. There are many types of shorelines depending on the geographical features. Sandy shores, rocky shores, and salt marshes are some examples of shorelines. Coastal areas are subject to tides, the periodic rising and lowering of the sea level. The rising tides push salt water inland, and the water covers parts of the coastline that are exposed during low tides. Because of the tides and the wave motion, animals and plants living along shorelines can tolerate high levels of salt and survive periodic submerging and exposure.

Temperate Oceans

Oceans are the largest habitats on earth. Oceans cover three-quarters of the surface of our planet and are home to a large variety of animals and plants. Oceans can be divided into three zones, based on the amount of light received.

The sunlit zone, closest to the surface, extends about 100 meters down. A large variety of microscopic floating algae live there and support the entire ocean food web. Many animals and plants also live in this area.

The second layer is the twilight zone. Very little light penetrates here, and therefore there are no plants. This layer is home to animals that can live without a lot of light.

Deeper down is the dark zone. This area is completely dark and very cold. This deep in the ocean, there is a lot of pressure from the water above. Very few organisms can live in these conditions.

Tropical Oceans

Tropical oceans are near the equator. They receive direct sunlight all year long and are very warm.

Tropical oceans are home to coral reefs. Coral reefs are structures built by a community of several thousand tiny organisms living together.

Terrestrial Biomes

Tundra

At the North and South Poles, the weather is very harsh. The tundra biome is characteristic of the polar climate. Temperatures are very cold year-round, and the soil is permanently frozen. There is very little **precipitation**, and the growing season is very short. There are no trees in the tundra. The vegetation consists of small plants called lichens and mosses. All animals and plants living in the tundra can live in very cold conditions.

Temperate Forests

The trees in temperate forests have large leaves that capture a lot of sunlight for photosynthesis. Trees shed their leaves in the fall, at the end of the growing season, and become **dormant** during winter. They regrow their leaves the following spring to restart photosynthesis. This way, trees adapt to the changing seasons.

Grasslands

Grasslands are big open spaces covered by grasses. There are very few trees and shrubs. Grasslands are found all over the world where average temperatures are mild and precipitation is moderate. They often occur between forests and deserts. The amount of rain precipitation determines the difference between an area being a grassland, a desert, or a forest. With a lot of precipitation, a grassland will become a forest. With less precipitation, it will become a desert.

Grasslands have rich soils suitable for agriculture. Often, large areas of grassland are used to grow crops.

Taiga (coniferous forests)

The land of the taiga is covered by vast forests. Winters are very cold, but the ground is not permanently frozen. Precipitation is also higher than in the tundra. These conditions allow trees to grow. The trees in the taiga are evergreen; they retain their leaves year round. The leaves are needle-like, and they are protected from cold by a waxy coating. Animals that inhabit the taiga can live in very cold weather.

Rainforests

There are two types of rainforests: tropical and temperate. Both types have lush vegetation and are very wet. Rainforests receive a lot of rain. The difference between tropical and temperate rainforests is in their average annual temperatures: tropical rainforests are warm, and temperate rainforests are cool.

Rainforests have an amazing variety of plants and animal species and are the most diverse habitats on land. Trees are very tall and form a green cover with their crowns, called the canopy. The canopy filters the light from the sun. The vegetation below is adapted to live in shaded conditions.

Deserts

Deserts are areas that receive little precipitation and experience extreme variations in temperature. Many deserts are very hot during the day and very cold during the night. Despite the extreme conditions of temperature and humidity, deserts are home to many different species of plants and animals.

Plants that live in this environment look very different from other plants. Because of the dry conditions, they have developed special adaptations to collect and store water. Shrubs are the dominant form of vegetation. Desert plants typically have very thick stems to store water and small or spiny leaves. They also have large root systems to collect the infrequent rainwater.

precipitation: water that falls to the ground as rain, snow, hail, or sleet.

dormant: asleep or inactive.

LIVING TOGETHER

Investigate

U.S. Biomes

So far you have learned about the main types of different world biomes and the climates where they are found. It is now time for you to work with your group to predict the types of biomes that might be found in the United States.

Use what you have learned about different biomes to prepare a list of the types of biomes you think might be part of the U.S. These questions can help you identify U.S. biomes.

- What types of aquatic biomes do you think are there? List all the aquatic biomes you think can be found in the U.S. Make sure to include both freshwater and marine biomes in your list.

- What types of terrestrial biomes do you think are there? Prepare a list of all terrestrial biomes in the U.S. Identify at least one place in the United States where you think a particular biome might exist and a reason for your thinking.

Communicate Your Ideas

Idea Briefing

On the blank transparency map of the U.S. provided by your teacher, draw the areas where you think each biome might occur. Use a blue marker to color aquatic biomes and a green marker to color terrestrial biomes. Label each biome. Be prepared to share your map with your class. Make sure you have a reason for including each biome and where it might occur.

The maps may not all be the same. As each group presents, look at where they placed their biomes and listen carefully to their reasoning. Discuss the evidence and reasoning each group is using, and try to agree, as a class, on where U.S. biomes are located.

Examine the map of U.S. biomes that your teacher will make available. How close were your predictions to the biomes scientists have identified? Was your class more accurate at identifying terrestrial biomes or aquatic biomes? Which were easier to identify? Discuss reasons for your answers

What are Adaptations?

Adaptations are special traits that help an organism survive in a given environment. They are characteristics or behaviors that are inherited by a plant or an animal. They are passed on from parent to offspring. Adaptations may be physical traits. Giraffes have long necks. Their long necks let them reach leaves of trees that are too high for other grazing animals to reach. Cacti have spiny "leaves." Water from plants with large leaves can escape through the leaves. The spines help cacti conserve water in hot and dry climates.

Adaptations may also be behaviors. Some animals, such as birds, migrate south when temperatures get cold. Other animals, such as bears, escape cold temperatures by hibernating.

In the competition for survival, organisms that have favorable adaptations have a greater chance of living longer and reproducing. Over many years, these desirable traits build up in a species. Unfavorable ones disappear. This process is called *natural selection*. The result is *evolution*. Evolution is a long, slow process.

adaptations: the special traits that allow an animal to survive in its environment.

LIVING TOGETHER

Answer the Big Question

Address the Big Challenge

You began this Unit by reading about a small town that needed your help. The town was faced with an important decision. You were asked to help them understand the possible results of different decisions they might make. As you worked through the Unit, you kept in mind the big question *How does water quality affect the ecology of a community?* and the advice you need to give the community. You've recorded claims and recommendations in the last column of the *Project Board*.

You are now ready to complete this Unit. Read about the town's situation one more time. Over the next few days, you will answer the big question as you address the challenge the town is facing.

Recall the Challenge

Wamego Needs Help!

Wamego (hwah-MEE-goh) is a small town with a population of about 1800. It is on the banks of the Crystal River. This town has always been a farming community. Most of the farmers grow corn and soybeans. These are the best crops to grow in this area. Nearly 95% of the residents are employed by Wamego's farming businesses. The local economy depends on farming. The other businesses in town all depend on the farmers and their employees (workers). These businesses include a grocery store, gas stations, a movie theater, and several restaurants.

The Crystal River is also important to Wamego. The river is a source of water for the crops. The river is also known as a good trout-fishing river.

Trout need very clean, cold water to thrive. Crystal River suits their needs. Every summer Wamego has a Trout Festival. Many people who enjoy fishing travel to the area. The festival celebrates trout fishing and preservation. The festival also educates people about what trout need to thrive. The goal of the education effort is to keep the number of trout at a healthy level. In that way, people can enjoy fishing there for many years to come. This festival is fun for many residents and tourists. It is also another income source for the residents of Wamego.

Lately, the farming business has not been good. Crop prices have dropped. The farmers are not making very much money. There is not enough to pay their workers or to support themselves. Some of the farmers have gone bankrupt. As a result, Wamego has lost 15% of its population during the last five years.

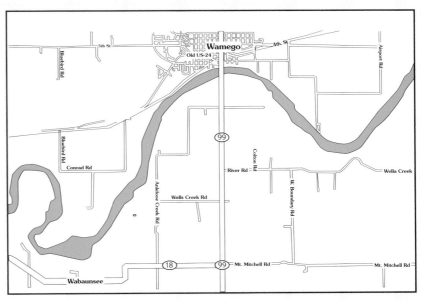

The town council is very concerned. They know farming will always be a part of life in Wamego. But they worry about the town losing too many people. They do not want to get so small that there will be very few businesses and residents in Wamego.

FabCo Wants to Move In

A mid-sized manufacturing company called FabCo has contacted the town council. FabCo manufactures cloth. The cloth is sold to companies that make clothes. FabCo is looking for a new location to build their company headquarters and manufacturing plant. FabCo is very interested in relocating to Wamego for several reasons.

- Wamego has a fairly large river and a train line running through the town. This, along with roads, would provide transportation routes for their products.

- The cost of living in the town is low. Their employees would like that.

- The river provides a natural resource (water). Water is important to the production of their cloth.

If FabCo is allowed to move to Wamego, the town could benefit as well. It would mean the following benefits:

- About 15,000 new residents would relocate to Wamego. This would require the building of many new homes, roads, and parks. A new school would need to be built. New businesses offering services to the company and the new residents would be needed. This means more buildings, parking lots, and roads would appear in Wamego.

- FabCo would offer many new jobs to Wamego's residents.

- The town would have money from taxes collected from FabCo and the new residents. This extra money could be used to improve life in Wamego in many ways, including a new hospital.

- The town would not have to depend on farming alone.

Sounds Great! So, What's the Problem?

Many of the residents, including some town council members, are concerned. They worry that FabCo's presence could mean problems for their community. Currently, the land is used for agriculture. If FabCo comes to town, the use of the land will change. The land will be needed for residential, commercial, and industrial purposes. Some people, including the organizers of the Trout Festival, wonder if this will change the river and the wildlife of Wamego.

Wamego residents are not the only ones concerned. Ten miles downstream is the town of St. George. It is also located along the Crystal River. St. George is an even smaller town than Wamego. It is a resort town. People travel from all over to vacation in St. George, using the river for recreation. There is fishing, swimming,

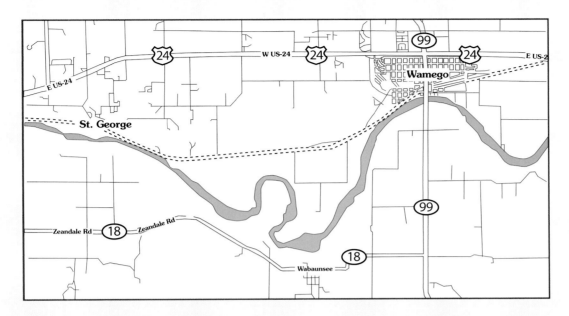

boating, hiking, and camping in the area. There are several hotels and bed & breakfasts that provide accommodations for tourists. The Crystal River's water quality is very important to St. George's economy and residents. The residents of St. George are worried that the changes in Wamego might affect their lives.

You have investigated the effects of changing land uses on water quality and the effect this can have on living things. You have learned how interconnected the different parts of an ecosystem are. To end this unit, your group will create a presentation to answer the big question for the town of Wamego.

Plan Your Answer

The town council knows that FabCo can bring money into the area. However, the council wants to know what to expect or what might happen if changes take place. The council is asking four people for their ideas.

- William Waters — organizer of the annual Wamego Trout Festival; fishing person
- Sara Song — FabCo executive; native of Wamego
- Ramone Ramirez — farmer
- Asha Adu — resort owner, Town of St. George

Your teacher will assign you to a new group. Each member of your group will be from one of the land-use groups you worked in during the Unit. Your group will represent one of these four people. You will prepare a presentation for the town council. Your presentation will focus on how FabCo's arrival could possibly change water quality from the point of view of the person you represent. You will need to think about why water quality is important to your assigned person. You will have to apply what you know about water quality and ecology to represent one of these individuals in your presentation.

Below: *a current map of the area*

Bottom: *map showing proposed changes*

Below is more information about how Wamego would change if FabCo arrived. On the next few pages is information on each of the individuals you are representing and their concerns. Use this information as you put together your argument.

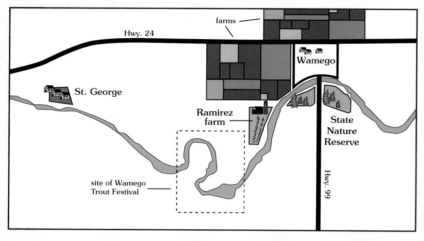

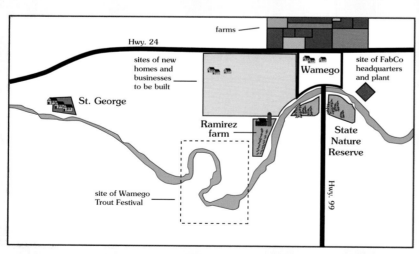

- The 15,000 new residents would need homes and apartment buildings to be built.

- New businesses and commercial areas would need to be developed to meet the needs of the population (restaurants, dry cleaners, day care centers, grocery stores, hospital, recreational facilities).

- New roads would be needed for the increased traffic to homes and businesses.

- All new storm drains for the new neighborhoods, businesses, parking lots, and streets would eventually drain into the Crystal River.

- Textile (cloth) manufacturing can often produce runoff (water) that can be acidic or basic and high in temperature.

- The size of Wamego's water-treatment plant could not handle all the new residents' needs. It would require updating. Until the upgrade is complete, the sewer system could have periodic overflows with raw sewage spilling into storm drains.

William Waters — organizer of the annual Wamego Trout Festival; fishing person

William has lived in Wamego his entire life. His family has been fishing the Crystal River for over 100 years. He works hard to protect the health of the Crystal River and the trout population in the river.

The Wamego Trout Festival location is downstream from the new plant location. William is worried about what might get into the stream to affect the trout population. He knows that trout are very sensitive to changes in the temperature, acidity, and dissolved oxygen in their water. Also, William is very concerned about all of the development exposing a lot of soil. He wonders what effect this condition would have on the quality of the stream and the plants in the stream.

Sara Song — FabCo executive; native of Wamego

Sara works for FabCo. Her current role for the company is to help relocate their headquarters and plant. Sara was born and raised in Wamego. She has not lived in Wamego for fifteen years, but she has been back to visit family often. Wamego has a very special place in Sara's heart. It was Sara's idea to move FabCo to Wamego. She was familiar with the town, and she knew that it would be a good match for the company. Sara also thinks that Wamego will benefit from this change without having to sacrifice its farming culture or its environment. Sara wants to deal with problems that could occur from

- the waste the plant's manufacturing process produces, and
- erosion at the site of the plant or near the new homes and businesses.

She wants the town council to know that she is aware of these issues, how they might be caused, and what problems they could cause. She will present this information along with some suggestions of how to prevent problems. Her goal is to ensure that both the town and the company benefit from this partnership.

Ramone Ramirez — farmer

Ramone owns and operates one of the largest soybean farms in the state. His farm is one of the few financially successful farms in the area. Ramone is very environmentally aware. He has been careful to ensure that his farm does not harm the Crystal River. Ramone has also successfully avoided the use of pesticides to keep herbivore insects from consuming his crops. Ramone uses predator insects to control the herbivore insect population.

Ramone irrigates (waters) his crops with water drawn from the Crystal River. The proposed site for new home and business development is next to Ramone's farm. The storm drains from these developments would enter the river upstream from his farm. He is very concerned about how changes in the condition of the water will affect his crops. Also, he wonders what changing the landscape will do to the predator insects he relies upon so much.

Asha Adu — resort owner, in the town of St. George

Asha runs a Bed & Breakfast in St. George, the town ten miles downstream from Wamego. Asha was born in Africa, but when she was still a baby, her family moved to St. George. Her parents opened the bed & breakfast when she was very young. Now she runs the business. Asha has spent almost her whole life growing up and living on the banks of the Crystal River. Asha is very concerned that the Crystal River will become polluted. She does not want to see the beauty of the place she has known and loved her whole life destroyed. She also does not want to have her tourism business affected by these changes.

Asha's brother works for the town of Wamego. He makes repairs and maintains the sewer system. Asha is aware of the condition and limits of the sewer system. She wonders what effect this could have on the water. Asha's tourists come to St.George because the river is so clear and clean. Asha wonders what changes could happen to the clarity and cleanliness of the river.

Communicate Your Ideas

Solution Showcase

The goal for a *Solution Showcase* is to have everyone better understand how a particular group approached their question or challenge. In this case, you get the opportunity to see the variety of ideas from different points of view. You will also have the chance to see the common ecological concepts important to everyone in the Wamego area. The *Solution Showcase* provides an opportunity for groups to share what they learned and how they have applied it to Wamego's problem.

Each group will present the argument of the person it represents and explain that person's ideas. Each group will discuss and formulate an argument based on its point of view and will present that argument in a *Solution Showcase*.

Each presentation should include several items to effectively communicate its individual's concerns and ideas. Each member of your group will be from a different one of the four land-use groups you have worked in during the Unit. This will allow you each to serve as an expert for each of the land uses. Each group member will have to remember and apply what they have learned during this Unit.

Each presentation should focus on the following items:

1. The water-quality issues your individual would be concerned about.

2. Sources of pollution your individual thinks FabCo would cause.

3. The problems your individual is concerned about. Consider the effect of these concerns on the organisms and food web for the area. Be very specific. The organisms living in this area are the same ones you examined in your food webs. Be sure to fully explain how the ecology problems you highlight would affect this food web.

4. Ideas your individual could suggest to reduce the harmful effects they are concerned about.

5. The water-quality tests that your individual would recommend if the changes took place. Be sure to describe the reason your individual thinks each test is needed.

You may want to use *Create Your Explanation* pages to help you think through and justify your claims and recommendations. You will probably need to refer to data from earlier investigations. The explanations and recommendations you and others made during the Unit might help you. Be sure to refer to the *Project Board* as a resource.

Your group will select one of three formats for your presentation.

Poster	You can construct a poster or posters. You will share the poster and its elements with the class. Your poster is not the presentation. Your poster is only a prop for you to refer to during your presentation. Your presentation should involve all members of your group.
PowerPoint®	You can construct a PowerPoint® presentation. Slides would have graphics and text that accompany your presentation. Avoid simply reading the slide to the class. Present your ideas using the slides as an aid. Your presentation should involve all members of your group.
Skit	You can create a skit. Your skit must have a written script that contains facts and concerns similar to what other groups will provide on their posters or PowerPoint presentations. Your skit should involve all members of your group.

Make sure to present the reasons you made the decisions you did. Your teacher will tell you how long you have to present. You will need to present your ideas quickly and clearly.

Project-Based Inquiry Science

Glossary

abiotic
nonliving.

abundance
a great amount (in this case, the number of a type of animal).

acid
a solution with a pH less than 7.

adaptations
the special traits that allow an animal to survive in its environment.

adapted to
suited for living in a given environment.

algae (singular, alga)
simple organisms that live in water. Some can be as small as one cell. Some are made up of many cells. These may be called "seaweed."

aquatic ecosystem
an ecosystem located in a body of water.

base
a solution with a pH greater than 7.

basicity
a term used to describe non-acid substances.

biotic
living.

classify
arrange or sort by categories.

community
groups of organisms living together in a certain area. The organisms interact and depend on each other for survival.

concentration
the amount of a substance mixed with another substance.

deposition
the setting down of Earth's materials onto another area.

dichotomous key
a key used to identify living things.

Glossary

diversity
difference (in this case, the different types of animals).

ecologist
a scientist who studies the relationships between organisms and their environment.

ecology
the study of how communities of plants, animals, and humans interact with each other and the physical environment.

ecosystem
all the living things in a given place, along with the nonliving environment.

elevation
the height of a geographical location above a reference point.

erosion
a process in which Earth's materials are loosened and removed.

groundwater
water that is located below the surface of the ground.

habitat
the place where an organism lives and grows naturally.

interaction
a kind of action in which two or more organisms have an effect on each other.

land use
how people use Earth's surface.

macroinvertebrate
an organism that does not have a backbone and can be seen with the naked eye.

manufacturing
the making or producing of anything.

microbe
an organism that cannot be seen with the unaided human eye. You need to use a microscope.

neutral
a solution with a pH of 7.

non-point-source pollution
pollution that comes from many sources over a large area.

pH
measure of how acidic a substance is.

pH indicator
a chemical that can be added to a solution to determine pH.

pH scale
a scale used by scientists to measure the acidity of a solution.

point-source pollution
pollution that originates from a single point or location.

pollution
substances added to air, water, or soil that cause harm to the environment.

raised-relief map
a three-dimensional map that shows elevations.

runoff

water from rain or melted snow that moves over the surface of the land.

simulate

to imitate how something happens in the real world by acting it out using a model.

slope

a measure of steepness. It is the ratio of the change in elevation to the change in horizontal distance (rise : run).

species

a group of organisms that look alike and can breed with each other and produce fertile offspring.

taxonomist

a scientist that classifies organisms by characteristics.

thermal pollution

a change in temperature in a natural body of water that is caused by humans.

turbidity

how cloudy, murky, or opaque something is (in this case, water).

turbulence

the violent disruption, agitation, or stirring up of something (in this case, of water).

watershed

the land area from which water drains into a particular stream, river, or lake.

Glosario

abiótico
referente a lo no vivo.

abundancia
una gran cantidad (en este caso, el número de tipos de animales).

ácido
una solución con pH menor que 7.

adaptaciones
los rasgos especiales que permiten que un animal sobreviva en su medio ambiente.

adaptado a
adecuado para vivir en un ambiente determinado.

agua residual
el agua de la lluvia o de nieve derretida que se dispersa sobre la superficie de la tierra.

agua subterránea
el agua que se localiza debajo de la superficie del suelo.

algas (singular, alga)
organismos simples que viven en el agua. Algunos están formados de muchas células. Se les pueden llamar "alga marina".

base
una solución con un pH mayor que 7.

basicidad
un término usado para describir sustancias no ácidas.

biótico
trelativo a lo vivo.

clasificar
organizar o clasificar por categorías.

clave dicotómica
una clave usada para identificar seres vivientes.

comunidad
los grupos de organismos que habitan de manera conjunta en un determinada área. Los organismos interactúan y dependen entre sí para sobrevivir.

concentración
la cantidad de un sustancia combinada con otra.

contaminación
sustancias agregadas al aire, agua o suelo que causan daño al medio ambiente.

contaminación termal
un cambio en la temperatura en un cuerpo natural de agua que es causada por los seres humanos.

cuenca
el área de tierra del cual el agua desemboca en un riachuelo, río o lago determinado.

deposición
asentamiento de los materiales de la tierra sobre otra área.

diversidad
diferencia (en este caso, la diferencia de tipos de animales).

ecología

el estudio de cómo las comunidades de plantas, animales y seres humanos interactúan entre sí y con el medio ambiente físico.

ecologista

un científico que estudia la relación entre los organismos y sus medios ambientes.

ecosistema

todos los seres vivos en un lugar determinado, junto con el ambiente no vivo.

ecosistema acuático

un ecosistema localizado en un cuerpo de agua.

elevación

la altura de una ubicación geográfica por encima de un punto de referencia.

erosión

un proceso por el cual los materiales de la tierra están aflojados y removidos.

escala pH

una escala que los científicos utilizan para medir la acidez de una solución.

especies

un grupo de organismos que son similares y pueden reproducirse entre ellos mismos y procrear crías fértiles.

fuente fija de contaminación

contaminación que se origina de un solo punto o localización.

hábitat

el lugar donde los organismos habitan y crecen de manera natural.

indicador de pH

un químico que se puede añadir a una solución para determinar su pH.

interacción

un tipo de acción en la que dos o más organismos se causan un efecto entre sí.

macro-invertebrado

un organismo que no tiene columna vertebral y se le puede ver a simple vista.

manufactura

la creación o producción de algo.

mapa de alto relieve

un mapa tridimensional que muestra elevaciones.

microbio

un organismo que no se puede ver a simple vista. Se necesita la ayuda de un microscopio.

neutral

una solución con un pH de 7.

pendiente

la medida de algo empinado. Es la proporción del cambio en elevación al cambio en la distancia horizontal.

Glosario

pH

la medida de cuán acida es una sustancia.

similar

imitar cómo sucede algo en una situación real al representarlo mediante un modelo.

solución de origen no puntual

contaminación que proviene de cualquier fuente hacia una área extensa.

taxonomista

un científico que clasifica los organismos de acuerdo a sus características.

turbiedad

cuán turbio, sucio u opaco es algo (el agua en este caso).

turbulencia

la violenta interrupción, agitación o turbulencia de algo (el agua en este caso).

uso del suelo

cómo las personas utilizan la superficie de la tierra

Picture Credits

Photos on pages 4, 23, left, 33, 63: *Jason Harris*

Photos on pages 6, lower left, 28, upper, 37, lower left: *continouswave.com*

Photo on page 11: *Wamego Recreation Department*

Photo on page 30: *Detroit News—Ricardo Thomas*

Photo on page 59: *Dane Deal*

Photo on page 124: *www.path.cam.ac.uk*

Photo on page 99: *wikipedia*

Photo on page 121: *Big Stock Photo*

Photos on pages 6, upper left, lower right, 7, 37, upper left, lower right, upper right, 38, right, 76, 139: *flickr*

Photos on pages 6, upper right, 7, upper left, 8, 9, lower, 10, 14, 17, 22, upper, 27, 31, 34, lower, 35, 36, 38, 45, 53, 56, 60, 62, 64, 65, lower, 70, lower, 73, 74, 75, 77, 80, 83, 84, 86, upper left, upper middle, upper right, lower, 91, 92, 95, 97, 98, 103, upper, 107, lower, 109, left, 110, upper left, upper right, 112, 113, left, second to left, second to right, right, 122, 123, upper left, upper middle, upper right, 124, left, right, 126, 127. 129, upper, 133, 135 middle, lower, 137, upper, middle, lower, 140: *istockphoto*

Photos on pages 22, lower, 134, upper, lower, 135, upper, middle, lower, 136, lower, 141, 142: *Fotolia*

All illustrations: *Dennis Falcon.*

All technical art: *Marie Killoran.*